Gödel's Proof

In 1931 Kurt Gödel published a revolutionary paper—one that challenged certain basic assumptions underlying much traditional research in mathematics and logic. Today his exploration of *terra incognita* has been recognized as one of the major contributions to modern scientific thought.

Here is the first book to present a readable explanation to both scholars and non-specialists of the main ideas, the broad implications of Gödel's proof. It offers any educated person with a taste for logic and philosophy the chance to gain genuine insight into a previously inaccessible subject.

In this new edition, Pulitzer prize–winning author Douglas R. Hofstadter has reviewed and updated the text of this classic work, clarifying ambiguities, making arguments clearer, and making the text more accessible. He has also added a new Foreword, which reveals his own unique personal connection to this seminal work and the impact it has had on his own professional life, explains the essence of Gödel's proof, and shows how and why Gödel's proof remains relevant today.

Gödel's

Proof

Revised Edition

by
Ernest Nagel
and
James R. Newman

Edited and with a New Foreword
by Douglas R. Hofstadter

New York University Press

New York and London

NEW YORK UNIVERSITY PRESS
New York and London

Library of Congress Cataloging-in-Publication Data
Nagel, Ernest, 1901–
Gödel's proof / by Ernest Nagel and James R. Newman.—Rev. ed. / edited
and with a new foreword by Douglas R. Hofstadter.
p. cm.
Includes bibliographical references and index.
ISBN 0-8147-5816-9 (acid-free paper)
1. Gödel's theorem. I. Newman, James Roy, 1907–1966. II. Hofstadter,
Douglas R., 1945– III. Title.
QA9.65 .N34 2002
511.3—dc21   2001044481

Manufactured in the United States of America
10 9 8 7 6 5 4 3 2 1

to
Bertrand Russell

Contents

Foreword to the New Edition

by Douglas R. Hofstadter

In August 1959, my family returned to Stanford, California, after a year in Geneva. I was fourteen, newly fluent in French, in love with languages, entranced by writing systems, symbols, and the mystery of meaning, and brimming with curiosity about mathematics and how the mind works.

One evening, my father and I went to a bookstore where I chanced upon a little book with the enigmatic title *Gödel's Proof.* Flipping through it, I saw many intriguing figures and formulas, and was particularly struck by a footnote about quotation marks, symbols, and symbols symbolizing other symbols. Intuitively sensing that *Gödel's Proof* and I were fated for each other, I knew I had to buy it.

As we walked out, my dad remarked that he had taken a philosophy course at City College of New York from one of its authors, Ernest Nagel, after which they had become good friends. This coincidence added to the book's mystique, and once home, I voraciously gobbled up its every word. From start to finish, *Gödel's Proof* resonated with my passions; suddenly I found myself obsessed with truth and falsity, paradoxes and proofs, mappings and mirrorings, symbol manipulation and symbolic logic, mathematics and metamathematics, the mystery of creative leaps in human thinking, and the mechanization of mentality.

Soon thereafter, my dad informed me that by chance he had run into Ernest Nagel on campus. Professor Nagel, normally at Columbia, happened to be spending a year at Stanford. Within a few days, our families got together, and I was charmed by all four Nagels—Ernest and Edith, and their two sons, Sandy and Bobby, both close to my age. I was thrilled to meet the author of a book I so loved, and I found Ernest and Edith to be enormously welcoming of my adolescent enthusiasms for science, philosophy, music, and art.

All too soon, the Nagels' sabbatical year had nearly drawn to a close, but before they left, they warmly invited me to spend a week that summer at their cabin in Vermont. During that idyllic stay, Ernest and Edith came to represent for me the acme of civility, generosity, and modesty; thus they remain in my memory, all these years later. The high point for me was a pair of sunny afternoons when Sandy and I sat outdoors in a verdant meadow and I read aloud to him the entirety of *Gödel's Proof.* What a twisty delight to read this book to the son of one of its authors!

By mail over the next few years, Sandy and I explored number patterns in a way that had a profound impact upon the rest of my life, and perhaps on his as well. He went on—known as Alex—to become a mathematics professor at the University of Wisconsin. Bobby, too, remained a friend and today he—known as Sidney—is a physics professor at the University of Chicago, and we see each other with great pleasure from time to time.

I wish I could say that I had met James Newman. I was given as a high-school graduation present his magnificent four-volume set, *The World of Mathematics,* and I always admired his writing style and his love for mathematics, but sad to say, we never crossed paths.

At Stanford I majored in mathematics, and my love for the ideas in Nagel and Newman's book inspired me to take a couple of courses in logic and metamathematics, but I was terribly disappointed by their aridity. Shortly thereafter, I entered graduate school in math and the same disillusionment recurred. I dropped out of math and turned to physics, but after a few years I found myself once again in a quagmire of abstractness and confusion.

One day in 1972, seeking some relief, I was browsing in the university bookstore and ran across *A Profile of Mathematical Logic* by Howard DeLong—a book that had nearly the same electrifying effect on me as *Gödel's Proof* did in 1959. This lucid treatise rekindled in me the long-dormant embers of my love for logic, metamathematics, and that wondrous tangle of issues I had connected with Gödel's theorem and its proof. Having long since lost my original copy of Nagel and Newman's magical booklet, I bought another one—luckily, it was still in print!—and reread it with renewed fascination.

That summer, taking a break from graduate school and driving across the continent, I camped out and religiously read about Gödel's work, the nature of reasoning, and the dream of mechanizing thought and consciousness. Without planning it, I wound up in New York City, and the first people I contacted were my old friends Ernest and Edith Nagel, who served as intellectual and emotional mentors for me. Over the next several months, I spent countless evenings in their apartment, and we ardently discussed many topics, including, of course, Gödel's proof and its ramifications.

The year 1972 marked the beginning of my own intense personal involvement with Gödel's theorem and the rich sphere of ideas surrounding it. Over the next few years, I developed an idiosyncratic set of explorations on this nexus of ideas, and wound up calling it *Gödel, Escher, Bach: an Eternal Golden Braid*. There is no doubt that the parents of my sprawling volume were Nagel and Newman's book, on the one hand, and Howard DeLong's book, on the other.

What is Gödel's work about? Kurt Gödel, an Austrian logician born in 1906, was steeped in the mathematical atmosphere of his time, which was characterized by a relentless drive toward formalization. People were convinced that mathematical thinking could be captured by laws of pure symbol manipulation. From a fixed set of axioms and a fixed set of typographical rules, one could shunt symbols around and produce new strings of symbols, called "theorems." The pinnacle of this movement was a monumental three-volume work by Bertrand Russell and Alfred North Whitehead called *Principia Mathematica,* which came out in the years 1910–1913. Russell and Whitehead believed that they had grounded all of mathematics in pure logic, and that their work would form the solid foundation for all of mathematics forevermore.

A couple of decades later, Gödel began to doubt this noble vision, and one day, while studying the extremely austere patterns of symbols in these volumes, he had a flash that those patterns were so much like number patterns that he could in fact replace each symbol by a number and reperceive all of *Principia Mathematica* not as symbol shunting but as number crunching (to borrow a modern term). This new way of looking at things had an astounding wraparound effect: since the subject matter of *Principia Mathematica* was numbers, and since Gödel had turned the medium of the volumes also into numbers, this showed that *Principia Mathematica* was its own subject matter, or in other words, that the patterned formulas of Russell and Whitehead's system could be seen as saying things about each other, or possibly even about themselves.

This wraparound was a truly unexpected turn of events, for it inevitably brought ancient paradoxes of self-reference to Gödel's mind—above all, "This statement is false." Using this type of paradox as his guide,

Gödel realized that, in principle, he could write down a formula of *Principia Mathematica* that perversely said about itself, "This formula is unprovable by the rules of *Principia Mathematica.*" The very existence of such a twisted formula was a huge threat to the edifice of Russell and Whitehead, for they had made the absolute elimination of "vicious circularity" a sacred goal, and had been convinced they had won the battle. But now it seemed that vicious circles had entered their pristine world through the back door, and Pandora's box was wide open.

The self-undermining Gödelian formula had to be dealt with, and Gödel did so most astutely, showing that although it resembled a paradox, it differed subtly from one. In particular, it was revealed to be a true statement that could not be proven using the rules of the system—indeed, a true statement whose unprovability resulted precisely from its truth.

In this shockingly bold manner, Gödel stormed the fortress of *Principia Mathematica* and brought it tumbling down in ruins. He also showed that his method applied to any system whatsoever that tried to accomplish the goals of *Principia Mathematica*. In effect, then, Gödel destroyed the hopes of those who believed that mathematical thinking is capturable by the rigidity of axiomatic systems, and he thereby forced mathematicians, logicians, and philosophers to explore the mysterious newly found chasm irrevocably separating provability from truth.

Ever since Gödel, it has been realized how subtle

and deep the art of mathematical thinking is, and the once-bright hope of mechanizing human mathematical thought starts to seem shaky, if not utterly quixotic. What, then, after Gödel, is mathematical thinking believed to be? What, after Gödel, is mathematical truth? Indeed, what is truth at all? These are the central issues that still lie unresolved, seventy years after Gödel's epoch-making paper appeared.

My book, despite owing a large debt to Nagel and Newman, does not agree with all of their philosophical conclusions, and here I would like to point out one key difference. In their "Concluding Reflections," Nagel and Newman argue that from Gödel's discoveries it follows that computers—"calculating machines," as they call them—are in principle incapable of reasoning as flexibly as we humans reason, a result that supposedly ensues from the fact that computers follow "a fixed set of directives" (i.e., a program). To Nagel and Newman, this notion corresponds to a fixed set of axioms and rules of inference—and the computer's behavior, as it executes its program, amounts to that of a machine systematically churning out proofs of theorems in a formal system. This mapping of computer onto formal system takes the term "calculating machine" very literally—that is, a machine built to deal with numbers and arithmetical facts alone. The idea that such machines by their very nature should churn out sets of true statements about mathematics is seductive and certainly has a grain of truth to it, but it is far from the full vision of the power and versatility of computers.

Although computers, as their name implies, are built of rigidly arithmetic-respecting hardware, nothing in their design links them inseparably to mathematical truth. It is no harder to get a computer to print out scads of false calculations ("2 + 2 = 5; 0/0 = 43," etc.) than to print out theorems in a formal system. A subtler challenge would be to devise "a fixed set of directives" by which a computer might explore the world of mathematical ideas (not just strings of mathematical symbols), guided by visual imagery, the associative patterns linking concepts, and the intuitive processes of guesswork, analogy, and esthetic choice that every mathematician uses.

When Nagel and Newman were composing *Gödel's Proof,* the goal of getting computers to think like people—in other words, artificial intelligence—was very new and its potential was unclear. The main thrust in those early days used computers as mechanical instantiations of axiomatic systems, and as such, they did nothing but churn out proofs of theorems. Now admittedly, if this approach represented the full scope of how computers might ever in principle be used to model cognition, then, indeed, Nagel and Newman would be wholly justified in arguing, based on Gödel's discoveries, that computers, no matter how rapid their calculations or how capacious their memories, are necessarily less flexible and insightful than the human mind.

But theorem-proving is among the least subtle of ways of trying to get computers to think. Consider the program "AM," written in the mid-1970s by Douglas Lenat. Instead of mathematical statements, AM dealt with concepts; its goal was to seek "interesting" ones, using a rudimentary model of esthetics and simplicity. Starting from scratch, AM discovered many concepts of number theory. Rather than logically proving theorems, AM wandered around the world of numbers, following its primitive esthetic nose, sniffing out patterns, and making guesses about them. As with a bright human, most of AM's guesses were right, some were wrong, and, for a few, the jury is still out.

For another way of modeling mental processes computationally, take neural nets—as far from the theorem-proving paradigm as one could imagine. Since the cells of the brain are wired together in certain patterns, and since one can imitate any such pattern in software— that is, in a "fixed set of directives"—a calculating engine's power can be harnessed to imitate microscopic brain circuitry and its behavior. Such models been studied now for many years by cognitive scientists, who have found that many patterns of human learning, including error making as an automatic by-product, are faithfully replicated.

The point of these two examples (and I could give many more) is that human thinking in all its flexible and fallible glory can in principle be modeled by a "fixed set of directives," provided one is liberated from the preconception that computers, built on arithmetical operations, can do nothing but slavishly produce truth, the whole truth, and nothing but the truth. That idea, admittedly, lies at the core of formal axiomatic reasoning systems, but today no one takes such systems seriously as a model of what the human mind does, even when it is at its most logical. We now understand that the human mind is fundamentally not a logic engine but an analogy engine, a learning engine, a guessing engine, an esthetics-driven engine, a self-correcting engine. And having profoundly understood this lesson, we are perfectly able to make "fixed sets of directives" that have some of these qualities.

To be sure, we have not yet come close to producing a computer program that has anything remotely resembling the flexibility of the human mind, and in this sense Ernest Nagel and James Newman were exactly on the mark in declaring, in their poetic fashion, that Gödel's theorem "is an occasion, not for dejection, but for a renewed appreciation of the powers of creative reason." It could not be said better.

There is, however, an irony to Nagel and Newman's interpretation of Gödel's result. Gödel's great stroke of genius—as readers of Nagel and Newman will see— was to realize that numbers are a universal medium for the embedding of patterns of any sort, and that for that reason, statements seemingly about numbers alone can in fact encode statements about other universes of discourse. In other words, Gödel saw beyond the surface level of number theory, realizing that numbers could represent any kind of structure. The analogous Gödelian leap with respect to computers would be to see that because computers at base manipulate numbers, and because numbers are a universal medium for the embedding of patterns of any sort, computers can deal with arbitrary patterns, whether they are logical or illogical, consistent or inconsistent. In short, when one steps back far enough from myriads of interrelated number patterns, one can make out patterns from other domains, just as the eye looking at a screen of pixels sees a familiar face and nary a 1 or 0. This Gödelian view of computers has exploded on the modern world to such an extent that today the numerical substrate of computers is all but invisible, except to specialists. Ordinary people routinely use computers for word processing, game playing, communication, animation, designing, drawing, and on and on, all without ever thinking about the basic arithmetical operations going on deep down in the hardware. Cognitive scientists, relying on the arithmetical hardware of their computers to be error-free and uncreative, give their computers "fixed sets of directives" to model human error-making and creativity. There is no reason to think that the processes of creative mathematical thinking cannot, at least in principle, be modeled using computers. But back in the 1950s, such visions of the potential of computers were hard to see. Still, it is ironic that in a book devoted to celebrating Gödel's insight that numbers engulf the world of patterns at large, the primary philosophical conclusion would be based on not heeding that insight, and would thereby fail to see that calculating machines can replicate patterns of any imaginable sort—even those of the creative human mind.

I shall close with a few words about why I have taken the liberty of making some technical emendations to this classic text. Although the book received mostly warm accolades from reviewers, there were some critics who felt that in spots it was not sufficiently precise and that it risked misleading its readers. The first time through, I myself was unaware of any such deficiencies, but many years later, when reading *Gödel's Proof* with an eye to explaining these same ideas myself as precisely and clearly as possible, I stumbled over certain passages in Chapter VII and realized, after a while, that the stumbling was not entirely my own fault. It made me sad to realize that this beloved book had a few blemishes, but there was obviously nothing I could do about it. Oddly enough, though, in the margins of my copy I carefully annotated all the glitches that I uncovered, indicating how they might be corrected—almost as if I had foreseen that one day I would receive an email out of the blue from New York University Press asking me if I would consider writing a foreword to a new edition of the book.

I must certainly be among the readers most profoundly affected by the little opus by Ernest Nagel and James Newman, and for that reason, having been given the chance, I owe it to them to polish their gem and to give it a new luster for the new millennium. I would like to believe that in so doing, I am not betraying my respected mentors but am instead paying them homage, as an ardent and faithful disciple.

Center for Research on Concepts and Cognition
Indiana University, Bloomington

Acknowledgments

The authors gratefully acknowledge the generous assistance they received from Professor John C. Cooley of Columbia University. He read critically an early draft of the manuscript, and helped to clarify the structure of the argument and to improve the exposition of points in logic. We wish to thank *Scientific American* for permission to reproduce several of the diagrams in the text, which appeared in an article on Gödel's Proof in the June 1956 issue of the magazine. We are indebted to Professor Morris Kline of New York University for helpful suggestions regarding the manuscript.

Gödel's Proof

I

Introduction

In 1931 there appeared in a German scientific periodical a relatively short paper with the forbidding title "Über formal unentscheidbare Sätze der Principia Mathematica und verwandter Systeme" ("On Formally Undecidable Propositions of Principia Mathematica and Related Systems"). Its author was Kurt Gödel, then a young mathematician of 25 at the University of Vienna and since 1938 a permanent member of the Institute for Advanced Study at Princeton. The paper is a milestone in the history of logic and mathematics. When Harvard University awarded Gödel an honorary degree in 1952, the citation described the work as one of the most important advances in logic in modern times.

At the time of its appearance, however, neither the title of Gödel's paper nor its content was intelligible to most mathematicians. The *Principia Mathematica* mentioned in the title is the monumental three-volume treatise by Alfred North Whitehead and Bertrand Russell on mathematical logic and the foundations of mathematics; and familiarity with that work is not a prerequisite to successful research in most branches of mathematics. Moreover, Gödel's paper deals with a set of questions that has never attracted more than a comparatively small group of students. The reasoning of the proof was so novel at the time of its publication that only those intimately conversant with the technical literature of a highly specialized field could follow the argument with ready comprehension. Nevertheless, the conclusions Gödel established are now widely recognized as being revolutionary in their broad philosophical import. It is the aim of the present essay to make the substance of Gödel's findings and the general character of his proof accessible to the nonspecialist.

Gödel's famous paper attacked a central problem in the foundations of mathematics. It will be helpful to give a brief preliminary account of the context in which the problem occurs. Everyone who has been exposed to elementary geometry will doubtless recall that it is taught as a *deductive* discipline. It is not presented as an experimental science whose theorems are to be accepted because they are in agreement with observation. This notion, that a proposition may be established as the conclusion of an explicit *logical proof*, goes back to the ancient Greeks, who discovered what is known as the "axiomatic method" and used it to develop geometry in a systematic fashion. The axiomatic method consists in accepting *without* proof certain propositions as axioms or postulates (e.g., the axiom that through two points just one straight line can be drawn), and then deriving from the axioms all other propositions of the system as theorems. The axioms constitute the "foundations" of the system; the theorems are the "superstructure," and are obtained from the axioms with the exclusive help of principles of logic.

The axiomatic development of geometry made a powerful impression upon thinkers throughout the ages; for the relatively small number of axioms carry the whole weight of the inexhaustibly numerous propositions derivable from them. Moreover, if in some way the truth of the axioms can be established—and, indeed, for some two thousand years most students believed without question that they are true of space— both the truth and the mutual consistency of all the theorems are automatically guaranteed. For these reasons the axiomatic form of geometry appeared to many generations of outstanding thinkers as the model of scientific knowledge at its best. It was natural to ask, therefore, whether other branches of thought besides geometry can be placed upon a secure axiomatic foundation. However, although certain parts of physics were given an axiomatic formulation in antiquity (e.g., by Archimedes), until modern times geometry was the only branch of mathematics that had what most students considered a sound axiomatic basis.

But within the past two centuries the axiomatic method has come to be exploited with increasing power and vigor. New as well as old branches of mathematics, including the study of the properties of the familiar cardinal (or "whole") numbers,* were supplied with what appeared to be adequate sets of axioms. A climate of opinion was thus generated in which it was tacitly assumed that each sector of mathematical thought can be supplied with a set of axioms sufficient for developing systematically the endless totality of true propositions about the given area of inquiry.

Gödel's paper showed that this assumption is untenable. He presented mathematicians with the astounding and melancholy conclusion that the axiomatic method has certain inherent limitations, which rule out the possibility that even the properties of the non-negative integers can ever be fully axiomatized. What

---

* Number theory is the study, going back to the ancient Greeks, of the properties of the natural numbers 0, 1, 2, 3, . . . —also sometimes called the "cardinal numbers" or "non-negative integers." Such properties include: how many factors a number has; how many different ways a number can be represented as a sum of smaller numbers; whether or not there is a biggest number having some specified property; whether or not certain equations have solutions that are whole numbers; and so on. Although number theory is inexhaustibly rich and full of surprises, its vocabulary is tiny—an alphabet of just a dozen symbols allows any number-theoretical statement to be expressed (although often cumbersomely).

In this book, we shall occasionally use the term "arithmetic" as a synonym for "number theory," but of course what this term entails is the full, rich universe of properties of the natural numbers, and not merely the mechanics of addition, subtraction, multiplication, and long division as taught in elementary schools, and as mechanized in cash registers and adding machines. [—*Ed.*]

is more, he proved that it is impossible to establish the internal logical consistency of a very large class of deductive systems—number theory, for example—unless one adopts principles of reasoning so complex that their internal consistency is as open to doubt as that of the systems themselves. In the light of these conclusions, no final systematization of many important areas of mathematics is attainable, and no absolutely impeccable guarantee can be given that many significant branches of mathematical thought are entirely free from internal contradiction.

Gödel's findings thus undermined deeply rooted preconceptions and demolished ancient hopes that were being freshly nourished by research on the foundations of mathematics. But his paper was not altogether negative. It introduced into the study of foundation questions a new technique of analysis comparable in its nature and fertility with the algebraic method that René Descartes introduced into geometry. This technique suggested and initiated new problems for logical and mathematical investigation. It provoked a reappraisal, still under way, of widely held philosophies of mathematics, and of philosophies of knowledge in general.

The details of Gödel's proofs in his epoch-making paper are too difficult to follow without considerable mathematical training. But the basic structure of his demonstrations and the core of his conclusions can be made intelligible to readers with very limited mathematical and logical preparation. To achieve such an understanding, the reader may find useful a brief account of certain relevant developments in the history of mathematics and of modern formal logic. The next four sections of this essay are devoted to this survey.

II

The Problem of Consistency

The nineteenth century witnessed a tremendous expansion and intensification of mathematical research. Many fundamental problems that had long withstood the best efforts of earlier thinkers were solved; new areas of mathematical study were created; and in various branches of the discipline new foundations were laid, or old ones entirely recast with the help of more precise techniques of analysis. To illustrate: the Greeks had proposed three problems in elementary geometry: with compass and straight-edge to trisect any angle, to construct a cube with a volume twice the volume of a given cube, and to construct a square equal in area to that of a given circle. For more than 2,000 years unsuccessful attempts were made to solve these problems; at last, in the nineteenth century it was proved that the desired constructions are logically impossible. There was, moreover, a valuable by-product of these labors. Since the solutions depend essentially upon determining the kind of roots that satisfy certain equations, concern with the celebrated exercises set in antiquity stimulated profound investigations into the nature of number and the structure of the number continuum. Rigorous definitions were eventually supplied for negative, complex, and irrational numbers; a logical basis was constructed for the real number system; and a new branch of mathematics, the theory of infinite numbers, was founded.

But perhaps the most significant development in its long-range effects upon subsequent mathematical history was the solution of another problem that the Greeks raised without answering. One of the axioms Euclid used in systematizing geometry has to do with parallels. The axiom he adopted is logically equivalent to (though not identical with) the assumption that through a point outside a given line only one parallel to the line can be drawn. For various reasons, this axiom did not appear "self-evident" to the ancients. They sought, therefore, to deduce it from the other Euclidean axioms, which they regarded as clearly self evident.[1] Can such a proof of the parallel axiom be

---

[1] The chief reason for this alleged lack of self-evidence seems to have been the fact that the parallel axiom makes an assertion about *infinitely remote* regions of space. Euclid defines parallel lines as straight lines in a plane that, "being produced indefinitely in both directions," do not meet. Accordingly, to say that two lines are parallel is to make the claim that the two lines will not meet even "at infinity." But the ancients were familiar with lines that, though they do not intersect each other in any finite region of the plane, do meet "at infinity." Such lines are said to be "asymptotic." Thus, a hyperbola is asymptotic to its axes. It was therefore not intuitively

given? Generations of mathematicians struggled with this question, without avail. But repeated failure to construct a proof does not mean that none can be found any more than repeated failure to find a cure for the common cold establishes beyond doubt that humanity will forever suffer from running noses. It was not until the nineteenth century, chiefly through the work of Gauss, Bolyai, Lobachevsky, and Riemann, that the *impossibility* of deducing the parallel axiom from the others was demonstrated. This outcome was of the greatest intellectual importance. In the first place, it called attention in a most impressive way to the fact that a *proof* can be given of the *impossibility of proving* certain propositions within a given system. As we shall see, Gödel's paper is a proof of the impossibility of formally demonstrating certain important propositions in number theory. In the second place, the resolution of the parallel axiom question forced the realization that Euclid is not the last word on the subject of geometry, since new systems of geometry can be constructed by using a number of axioms different from, and incompatible with, those adopted by Euclid. In particular, as is well known, immensely interesting and fruitful results are obtained when Euclid's parallel axiom is replaced by the assumption that more than one parallel can be drawn to a given line through a given point, or, alter-

---

evident to the ancient geometers that from a point outside a given straight line only one straight line can be drawn that will not meet the given line even at infinity.

natively, by the assumption that no parallels can be drawn. The traditional belief that the axioms of geometry (or, for that matter, the axioms of any discipline) can be established by their apparent self-evidence was thus radically undermined. Moreover, it gradually became clear that the proper business of pure mathematicians is to *derive theorems from postulated assumptions,* and that it is not their concern whether the axioms assumed are actually true. And, finally, these successful modifications of orthodox geometry stimulated the revision and completion of the axiomatic bases for many other mathematical systems. Axiomatic foundations were eventually supplied for fields of inquiry that had hitherto been cultivated only in a more or less intuitive manner. (See Appendix, no. 1.)

The over-all conclusion that emerged from these critical studies of the foundations of mathematics is that the age-old conception of mathematics as "the science of quantity" is both inadequate and misleading. For it became evident that mathematics is simply the discipline *par excellence* that draws the conclusions logically implied by any given set of axioms or postulates. In fact, it came to be acknowledged that the validity of a mathematical inference in no sense depends upon any special meaning that may be associated with the terms or expressions contained in the postulates. Mathematics was thus recognized to be much more abstract and formal than had been traditionally supposed: more abstract, because mathematical statements can be construed in principle to be about anything what-

soever rather than about some inherently circum-
scribed set of objects or traits of objects; and more
formal, because the validity of mathematical demonstrations is grounded in the structure of statements,
rather than in the nature of a particular subject matter.
The postulates of any branch of demonstrative mathematics are not inherently about space, quantity, apples,
angles, or budgets; and any special meaning that may
be associated with the terms (or "descriptive predicates") in the postulates plays no essential role in the
process of deriving theorems. We repeat that the sole
question confronting the pure mathematician (as distinct from the scientist who employs mathematics in
investigating a special subject matter) is not whether
the postulates assumed or the conclusions deduced
from them are true, but whether the alleged conclusions are in fact the *necessary logical consequences* of the
initial assumptions.

Take this example. Among the undefined (or "primitive") terms employed by the influential German
mathematician David Hilbert in his famous axiomati-
zation of geometry (first published in 1899) are
'point', 'line', 'lies on', and 'between'. We may grant
that the customary meanings connected with these expressions play a role in the process of discovering and
learning theorems. Since the meanings are familiar, we
feel we understand their various interrelations, and
they motivate the formulation and selection of axioms;
moreover, they suggest and facilitate the formulation
of the statements we hope to establish as theorems.

Yet, as Hilbert plainly states, insofar as we are concerned with the primary mathematical task of exploring the purely logical relations of dependence between statements, the familiar connotations of the primitive terms are to be ignored, and the sole "meanings" that are to be associated with them are those assigned by the axioms into which they enter.[2] This is the point of Russell's famous epigram: pure mathematics is the subject in which we do not know what we are talking about, or whether what we are saying is true.

A land of rigorous abstraction, empty of all familiar landmarks, is certainly not easy to get around in. But it offers compensations in the form of a new freedom of movement and fresh vistas. The intensified formalization of mathematics emancipated people's minds from the restrictions that the customary interpretation of expressions placed on the construction of novel systems of postulates. New kinds of algebras and geometries were developed which marked significant departures from the mathematics of tradition. As the meanings of certain terms became more general, their use became broader and the inferences that could be drawn from them less confined. Formalization led to a great variety of systems of considerable mathematical interest and value. Some of these systems, it must be

---

[2] In more technical language, the primitive terms are "implicitly" defined by the axioms, and whatever is not covered by the implicit definitions is irrelevant to the demonstration of theorems.

admitted, did not lend themselves to interpretations as obviously intuitive (i.e., commonsensical) as those of Euclidean geometry or arithmetic, but this fact caused no alarm. Intuition, for one thing, is an elastic faculty: our children will probably have no difficulty in accepting as intuitively obvious the paradoxes of relativity, just as we do not boggle at ideas that were regarded as wholly unintuitive a couple of generations ago. Moreover, as we all know, intuition is not a safe guide: it cannot properly be used as a criterion of either truth or fruitfulness in scientific explorations.

However, the increased abstractness of mathematics raised a more serious problem. It turned on the question whether a given set of postulates serving as foundation of a system is internally consistent, so that no mutually contradictory theorems can be deduced from the postulates. The problem does not seem pressing when a set of axioms is taken to be about a definite and familiar domain of objects; for then it is not only significant to ask, but it may be possible to ascertain, whether the axioms are indeed true of these objects. Since the Euclidean axioms were generally supposed to be true statements about space (or objects in space), no mathematician prior to the nineteenth century ever considered the question whether a pair of contradictory theorems might some day be deduced from the axioms. The basis for this confidence in the consistency of Euclidean geometry is the sound principle that logically incompatible statements cannot be simultaneously true; accordingly, if a set of statements is true (and this was assumed of the Euclidean axioms), these statements are mutually consistent.

The non-Euclidean geometries were clearly in a different category. Their axioms were initially regarded as being plainly false of space, and, for that matter, doubtfully true of anything; thus the problem of establishing the internal consistency of non-Euclidean systems was recognized to be both formidable and critical. In elliptic geometry, for example, Euclid's parallel postulate is replaced by the assumption that through a given point outside a line *no* parallel to it can be drawn. Now consider the question: Is the elliptic set of postulates consistent? The postulates are apparently not true of the space of ordinary experience. How, then, is their consistency to be shown? How can one prove they will not lead to contradictory theorems? Obviously the question is not settled by the fact that the theorems already deduced do not contradict each other—for the possibility remains that the very next theorem to be deduced may upset the apple cart. But, until the question is settled, one cannot be certain that elliptic geometry is a true alternative to the Euclidean system, i.e., equally valid mathematically. The very possibility of non-Euclidean geometries was thus contingent on the resolution of this problem.

A general method for solving it was devised. The underlying idea is to find a "model" (or "interpretation") for the abstract postulates of a system, so that each postulate is converted into a true statement about the model. In the case of Euclidean geometry, as we have noted, the model was ordinary space. The method was used to find other models, the elements of which could serve as crutches for determining the consistency of abstract postulates. The procedure goes something like this. Let us understand by the word 'class' a collection or aggregate of distinguishable elements, each of which is called a 'member' of the class. Thus, the class of prime numbers less than 10 is the collection whose members are 2, 3, 5, and 7. Suppose the following set of postulates concerning two classes K and L, whose special nature is left undetermined except as "implicitly" defined by the postulates:

1. Any two members of K are contained in just one member of L.

2. No member of K is contained in more than two members of L.

3. The members of K are not all contained in a single member of L.

4. Any two members of L contain just one member of K.

5. No member of L contains more than two members of K.

From this small set we can derive, by using customary rules of inference, a number of theorems. For example, it can be shown that K contains just three members. But is the set consistent, so that mutually contradictory theorems can never be derived from it? The question can be answered readily with the help of the following model:

> Let K be the class of points consisting of the vertices of a triangle, and L the class of lines made up of its sides; and let us understand the phrase 'a member of K is contained in a member of L' to mean that a point which is a vertex lies on a line which is a side. Each of the five abstract postulates is then converted into a true statement. For instance, the first postulate asserts that any two points which are vertices of the triangle lie on just one line which is a side. (See Fig. 1.) In this way the set of postulates is proved to be consistent.

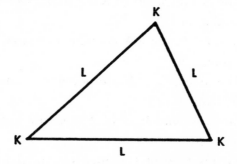

Fig. 1

Model for a set of postulates about two classes, K and L, is a triangle whose vertices are the members of K and whose sides are the members of L. The geometrical model shows that the postulates are consistent.

The consistency of plane elliptic geometry can also, ostensibly, be established by a model embodying the postulates. We may interpret the expression 'plane' in the axioms of elliptic geometry as the surface of a Euclidean sphere, the expression 'point' as a pair of antipodal points on this surface, the expression 'straight line' as a great circle on this surface, and so on. Each elliptic postulate is then converted into a theorem of Euclid. For example, on this interpretation the elliptic parallel postulate reads: Through a point on the surface of a sphere, no great circle can be drawn parallel to a given great circle. (See Fig. 2.)

At first glance this proof of the consistency of elliptic geometry may seem conclusive. But a closer look is disconcerting. For a sharp eye will discern that the problem has not been solved; it has merely been shifted to another domain. The proof attempts to settle the consistency of elliptic geometry by appealing to the consistency of Euclidean geometry. What emerges, then, is only this: elliptic geometry is consistent if Euclidean geometry is consistent. The authority of Euclid is thus invoked to demonstrate the consistency of a system which challenges the exclusive validity of Euclid. The inescapable question is: Are the axioms of the Euclidean system itself consistent?

An answer to this question, hallowed, as we have noted, by a long tradition, is that the Euclidean axioms are true and are therefore consistent. This answer is no longer regarded as acceptable; we shall return to it presently and explain why it is unsatisfactory. Another answer is that the axioms jibe with our actual, though limited, experience of space and that we are justified

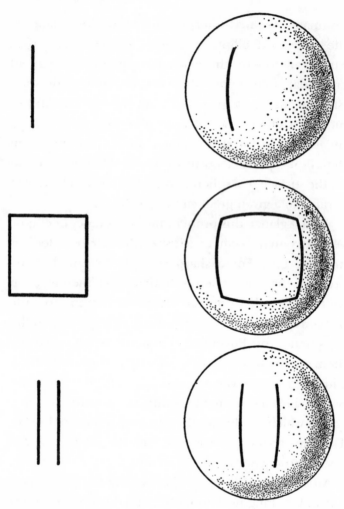

Fig. 2

The non-Euclidean geometry of the "elliptic plane" can be represented by a Euclidean model. The elliptic plane becomes the surface of a Euclidean sphere, points on the plane become pairs of antipodal points on this surface, straight lines in the plane become great circles. Thus, a portion of the elliptic plane bounded by in extrapolating from the small to the universal. But, although much inductive evidence can be adduced to support this claim, our best proof would be logically incomplete. For even if all the observed facts are in agreement with the axioms, the possibility is open that a hitherto unobserved fact may contradict them and so destroy their title to universality. Inductive considerations can show no more than that the axioms are plausible or probably true.

Hilbert tried yet another route to the top. The clue to his way lay in Cartesian coordinate geometry. In his interpretation Euclid's axioms were simply transformed into algebraic truths. For instance, in the axioms for plane geometry, construe the expression 'point' to signify a pair of numbers, the expression 'straight line' the (linear) relation between numbers expressed by a first degree equation with two unknowns, the expression 'circle' the relation between numbers expressed by a quadratic equation of a certain form, and so on. The geometric statement that two distinct points uniquely determine a straight line is then transformed into the algebraic truth that two distinct pairs of numbers uniquely determine a linear relation; the geometric theorem that a straight line intersects a circle in at most two points, into the algebraic

segments of straight lines is depicted as a portion of the sphere bounded by parts of great circles (*center*). Two line segments in the elliptic plane are two segments of great circles on the Euclidean sphere (*bottom*), and these, if extended, indeed intersect, thus contradicting the Euclidean parallel postulate.

theorem that a pair of simultaneous equations in two unknowns (one of which is linear and the other quadratic of a certain type) determine at most two pairs of real numbers; and so on. In brief, the consistency of the Euclidean postulates is established by showing that they are satisfied by an algebraic model. This method of establishing consistency is powerful and effective. Yet it, too, is vulnerable to the objection already set forth. For, again, a problem in one domain is resolved by transferring it to another. Hilbert's argument for the consistency of his geometric postulates shows that if algebra is consistent, so is his geometric system. The proof is clearly relative to the assumed consistency of another system and is not an "absolute" proof.

In the various attempts to solve the problem of consistency there is one persistent source of difficulty. It lies in the fact that the axioms are interpreted by models composed of an infinite number of elements. This makes it impossible to encompass the models in a finite number of observations; hence the truth of the axioms themselves is subject to doubt. In the inductive argument for the truth of Euclidean geometry, a finite number of observed facts about space are presumably in agreement with the axioms. But the conclusion that the argument seeks to establish involves an extrapolation from a finite to an infinite set of data. How can we justify this jump? On the other hand, the difficulty is minimized, if not completely eliminated, where an appropriate model can be devised that contains only a finite number of elements. The triangle model used to show the consistency of the five abstract postulates for the classes K and L is finite; and it is comparatively simple to determine by actual inspection whether all the elements in the model actually satisfy the postulates, and thus whether they are true (and hence consistent). To illustrate: by examining in turn all the vertices of the model triangle, one can learn whether any two of them lie on just one side—so that the first postulate is established as true. Since all the elements of the model, as well as the relevant relations among them, are open to direct and exhaustive inspection, and since the likelihood of mistakes occurring in inspecting them is practically nil, the consistency of the postulates in this case is not a matter for genuine doubt.

Unfortunately, most of the postulate systems that constitute the foundations of important branches of mathematics cannot be mirrored in finite models. Consider the postulate in elementary arithmetic which asserts that every integer has an immediate successor differing from any preceding integer. It is evident that the model needed to test the set to which this postulate belongs cannot be finite, but must contain an infinity of elements. It follows that the truth (and so the consistency) of the set cannot be established by an exhaustive inspection of a limited number of elements. Apparently we have reached an impasse. Finite models suffice, in principle, to establish the consistency of certain sets of postulates; but these are of slight mathematical importance. Non-finite models, necessary for the interpretation of most postulate systems of mathematical significance, can be described only in general terms; and we cannot conclude as a matter of course that the descriptions are free from concealed contradictions.

It is tempting to suggest at this point that we can be sure of the consistency of formulations in which non-finite models are described if the basic notions employed are transparently "clear" and "distinct." But the history of thought has not dealt kindly with the doctrine of clear and distinct ideas, or with the doctrine of intuitive knowledge implicit in the suggestion. In certain areas of mathematical research in which assumptions about infinite collections play central roles, radical contradictions have turned up, in spite of the intuitive clarity of the notions involved in the assumptions and despite the seemingly consistent character of the intellectual constructions performed. Such contradictions (technically referred to as "antinomies") have emerged in the theory of infinite numbers, developed by Georg Cantor in the nineteenth century; and the occurrence of these contradictions has made plain that the apparent clarity of even such an elementary notion as that of *class** (or *aggregate*) does not guarantee the

---

* In this book, we use the term "class" to mean what most people today tend to call "sets." In the decades when Whitehead and Russell developed *Principia Mathematica* and Gödel made his discoveries, the more common term was "class," however, and accordingly, we shall use that word, since it is more reflective of the epoch of which we are writing. [—*Ed.*]

consistency of any particular system built on it. Since the mathematical theory of classes, which deals with the properties and relations of aggregates or collections of elements, is often adopted as the foundation for other branches of mathematics, and in particular for number theory, it is pertinent to ask whether contradictions similar to those encountered in the theory of infinite classes infect the formulations of other parts of mathematics.

In point of fact, Bertrand Russell constructed a contradiction within the framework of elementary logic itself that is precisely analogous to the contradiction first developed in the Cantorian theory of infinite classes. Russell's antinomy can be stated as follows. Classes seem to be of two kinds: those which do not contain themselves as members, and those which do. A class will be called "normal" if, and only if, it does not contain itself as a member; otherwise it will be called "nonnormal." An example of a normal class is the class of mathematicians, for patently the class itself is not a mathematician and is therefore not a member of itself. An example of a non-normal class is the class of all thinkable things; for the class of all thinkable things is itself thinkable and is therefore a member of itself. Let 'N' by definition stand for the class of *all* normal classes. We ask whether N itself is a normal class. If N is normal, it is a member of itself (for by definition N contains all normal classes); but, in that case, N is nonnormal, because by definition a class that contains itself as a member is non-normal. On the other hand, if N is non-normal, it is a member of itself (by definition of "non-normal"); but, in that case, N is normal, because by definition the members of N are normal classes. In short, N is normal if, and only if, N is non-normal. It follows that the statement 'N is normal' is both true and false. This fatal contradiction results from an uncritical use of the apparently pellucid notion of "class." Other paradoxes were found later, each of them constructed by means of familiar and seemingly cogent modes of reasoning. Mathematicians came to realize that in developing consistent systems familiarity and intuitive clarity are weak reeds to lean on.

We have seen the importance of the problem of consistency, and we have acquainted ourselves with the classically standard method for solving it with the help of models. It has been shown that in most instances the problem requires the use of a non-finite model, the description of which may itself conceal inconsistencies. We must conclude that, while the model method is an invaluable mathematical tool, it does not supply a final answer to the problem it was designed to solve.

III

Absolute Proofs of Consistency

The limitations inherent in the use of models for establishing consistency, and the growing apprehension that the standard formulations of many mathematical systems might all harbor internal contradictions, led to new attacks upon the problem. An alternative to relative proofs of consistency was proposed by Hilbert. He sought to construct "absolute" proofs, by which the consistency of systems could be established without assuming the consistency of some other system. We must briefly explain this approach as a further preparation for understanding Gödel's achievement.

The first step in the construction of an absolute proof, as Hilbert conceived the matter, is the *complete formalization* of a deductive system. This involves draining the expressions occurring within the system of all meaning: they are to be regarded simply as empty signs. How these signs are to be combined and manipulated is to be set forth in a set of precisely stated rules. The purpose of this procedure is to construct a system of signs (called a "calculus") which conceals nothing

and which has in it only that which we explicitly put
into it. The postulates and theorems of a completely
formalized system are "strings" (or finitely long sequences) of *meaningless marks,* constructed according
to rules for combining the elementary signs of the
system into larger wholes. Moreover, when a system
has been completely formalized, the derivation of
theorems from postulates is nothing more than the
transformation (pursuant to rule) of one set of such
"strings" into another set of "strings." In this way the
danger is eliminated of using any unavowed principles of reasoning. Formalization is a difficult and
tricky business, but it serves a valuable purpose. It reveals structure and function in naked clarity, as does
a cutaway working model of a machine. When a system has been formalized, the logical relations between mathematical propositions are exposed to view;
one is able to see the structural patterns of various
"strings" of "meaningless" signs, how they hang together, how they are combined, how they nest in one
another, and so on.

A page covered with the "meaningless" marks of
such a formalized mathematics does not *assert* anything
—it is simply an abstract design or a mosaic possessing
a determinate structure.* Yet it is clearly possible to

---

* A more accurate way to describe such a formalized calculus is
to say that its symbols may *appear* to have meanings (and its rules
may well push its symbols around in such a way that they *act* exactly
as symbols with the desired meanings would act), but their behavior

describe the configurations of such a system and to make statements about the configurations and their various relations to one another. One may say that a "string" is palindromic (i.e., is the same when read backwards as forwards) or that it has two symbols in common with another "string," or that one "string" is made up of three others, and so on. Such statements are evidently meaningful and may convey important information about the formal system. It must now be observed, however, that such meaningful statements *about* a meaningless (or formalized) mathematical system plainly do not themselves belong to that system. They belong to what Hilbert called "meta-mathematics," to the language that is *about* mathematics. Meta-mathematical

---

is not a *consequence* of their meanings; indeed, quite the reverse is the case. To the extent that formal symbols appear meaningful, this appearance devolves entirely from their behavior, which in turn is wholly determined by the system's rules and initial formulas (axioms). It is therefore not entirely unjustified or unreasonable to see "meaningless strings" as having a type of meaningfulness, as long as one bears in mind that any such meaning is *passive* rather than active. To explain this notion metaphorically, the strings and their constituent symbols couldn't care less about any putative meanings that anybody might wish to read into them—all that matters to them is how the rules manipulate them.

To give an analogy, you might give your car a pet name and even think of it as animate, but the car will function exactly the same with or without its pet name and its putative "soul"; all that matters is the machinery that makes it work. The name and soul have no effect on that machinery, although they may make it easier for you to relate to your car. As with favorite cars, so with formal calculi. [—*Ed.*]

statements are statements about the signs occurring within a formalized mathematical system (i.e., a calculus)—about the kinds and arrangements of such signs when they are combined to form longer strings of marks called "formulas," or about the relations between formulas that may obtain as a consequence of the rules of manipulation specified for them.

A few examples will help to convey an understanding of Hilbert's distinction between mathematics (i.e., a system of meaningless signs) and meta-mathematics (meaningful statements about mathematics, the signs occurring in the calculus, their arrangement and relations). Consider the expression:

$$2 + 3 = 5$$

This expression belongs to mathematics (arithmetic) and is constructed entirely out of elementary arithmetical signs. On the other hand, the *statement*

'$2 + 3 = 5$' is an arithmetical formula

asserts something about the displayed expression. The statement does not express an arithmetical fact and does not belong to the formal language of arithmetic; it belong to meta-mathematics, because it characterizes a certain string of arithmetical signs as being a formula. The following statement belongs to meta-mathematics:

If the sign '$=$' is to be used in a formula of arithmetic, the sign must be flanked both left and right by numerical expressions.

This statement lays down a necessary condition for using a certain arithmetical sign in arithmetical formulas: the structure that an arithmetical formula must have if it is to embody that sign.

Consider next the three formulas:

$$x = x$$
$$0 = 0$$
$$0 \neq 0$$

Each of these belongs to mathematics, because each is built up entirely out of mathematical signs. But the statement:

'x' is a variable

belongs to meta-mathematics, since it characterizes a certain mathematical sign as belonging to a specific class of signs (i.e., to the class of variables). Again, the following statement belongs to meta-mathematics:

The formula '$0 = 0$' is derivable from the formula '$x = x$' by substituting the numeral '0' for the variable 'x'.

It specifies in what manner one mathematical formula can be obtained from another formula, and thereby describes how the two formulas are related to each other. Similarly, the statement

'$0 \neq 0$' is not a theorem of formal system X

belongs to meta-mathematics, for it says of a certain formula that it is not derivable from the axioms of the specific formal calculus mentioned, and thus asserts that a certain relation does not hold between the indicated formulas of the system. Finally, the next statement belongs to meta-mathematics:

Formal system X is consistent

(i.e., it is not possible to derive from the axioms of system X two formally contradictory formulas—for example, the formulas '0 = 0' and '0 ≠ 0'). This is patently about a formal calculus, and asserts that pairs of formulas of a certain sort do not stand in a specific relation to the formulas that constitute the axioms of that calculus.[3]

---

[3] It is worth noting that the meta-mathematical statements given in the text do not contain as constituent parts of themselves any of the *mathematical signs and formulas* that appear in the examples. At first glance this assertion seems palpably untrue, for the signs and formulas are plainly visible. But, if the statements are examined with an analytic eye, it will be seen that the point is well taken. The meta-mathematical statements contain the *names* of certain arithmetical expressions, but not the arithmetical expressions themselves. The distinction is subtle but both valid and important. It arises out of the circumstance that the rules of English grammar require that no sentence literally contain the objects to which the expressions in the sentence may refer, but only the *names* of such objects. Obviously, when we talk about a city we do not put the city itself into a sentence, but only the name of the city; and, similarly, if we wish to say something about a word (or other linguistic sign), it is not the word itself (or the sign) that can appear in the sentence, but only a name for the word (or sign). According to a

It may be that the reader finds the word 'meta-mathematics' ponderous and the concept puzzling. We shall not argue that the word is pretty; but the concept itself will perplex no one if we point out that it is used in connection with a special case of a well-known distinction, namely between a subject matter under study and discourse about the subject matter. The statement 'among phalaropes the males incubate the eggs' pertains to the subject matter investigated by zoologists, and belongs to zoology; but if we say that this assertion about phalaropes proves that zoology is irrational, our statement is not about phalaropes, but about the assertion and the discipline in which it occurs, and is meta-zoology. If we say that the *id* is mightier than the *ego,* we are making noises that belong to orthodox psychoanalysis; but if we criticize this statement as meaning-

---

standard convention we construct a name for a linguistic expression by placing single quotation marks around it. Our text adheres to this convention. It is correct to write:

Chicago is a populous city.

But it is incorrect to write:

Chicago is tri-syllabic.

To express what is intended by this latter sentence, one must write:

'Chicago' is tri-syllabic.

Likewise, it is incorrect to write:

$x = 5$ is an equation.

We must, instead, formulate our intent by:

'$x = 5$' is an equation.

less and unprovable, our criticism belongs to meta-psychoanalysis. And so in the case of mathematics and meta-mathematics. The formal systems that mathematicians construct belong in the file labeled "mathematics"; the description, discussion, and theorizing about the systems belong in the file marked "meta-mathematics."

The importance to our subject of recognizing the distinction between mathematics and meta-mathematics cannot be overemphasized. Failure to respect it has produced paradoxes and confusion. Recognition of its significance has made it possible to exhibit in a clear light the logical structure of mathematical reasoning. The merit of the distinction is that it entails a careful codification of the various signs that go into the making of a formal calculus, free of concealed assumptions and possibly misleading associations of meaning. Furthermore, it requires exact definitions of the operations and logical rules of mathematical construction and deduction, many of which mathematicians had applied without being explicitly aware of what they were using.

Hilbert saw to the heart of the matter, and it was upon the distinction between a formal calculus and its description that he based his attempt to build "absolute" proofs of consistency. Specifically, he sought to develop a method that would yield demonstrations of consistency as much beyond genuine logical doubt as the use of finite models for establishing the consistency of certain sets of postulates—by an analysis of a finite number of structural features of expressions in completely formalized calculi. The analysis consists in noting the various types of signs that occur in a calculus, indicating how to combine them into formulas, prescribing how formulas can be obtained from other formulas, and determining whether formulas of a given kind are derivable from others through explicitly stated rules of operation. Hilbert believed it might be possible to exhibit every mathematical calculus as a sort of "geometrical" pattern of formulas, in which the formulas stand to each other in a finite number of structural relations. He therefore hoped to show, by exhaustively examining these structural properties of expressions within a system, that formally contradictory formulas cannot be obtained from the axioms of given calculi. An essential requirement of Hilbert's program in its original conception was that demonstrations of consistency involve only such procedures as make no reference either to an infinite number of structural properties of formulas or to an infinite number of operations with formulas. Such procedures are called "finitistic"; and a proof of consistency conforming to this requirement is called "absolute." An "absolute" proof achieves its objectives by using a minimum of principles of inference, and does not assume the consistency of some other set of axioms. An absolute proof of the consistency of a formalized version of number theory, if such a proof could be constructed, would therefore show by a finitistic meta-mathematical procedure that two contradictory formulas, such as '0 = 0' and its formal negation '$\sim (0 = 0)$'—where the sign '$\sim$' is used in a rule-bound way so as to mimic, formally, our intuitive concept of "negation"—cannot both be derived by stated rules of inference from the axioms (or initial formulas).[4]

It may be useful, by way of illustration, to compare meta-mathematics as a theory of proof with the theory of chess. Chess is played with 32 pieces of specified design on a square board containing 64 square subdivisions, where the pieces may be moved in accordance with fixed rules. The game can obviously be played without assigning any "interpretation" to the pieces or to their various positions on the board, although such an interpretation could be supplied if desired. For example, we could stipulate that a given pawn is to represent a certain regiment in an army, that a given square is to stand for a certain geographical region, and so on. But such stipulations (or interpretations) are not customary; and neither the pieces, nor the squares, nor the positions of the pieces on the board signify anything *outside* the game. In this sense, the pieces and their configurations on the board are "meaningless." Thus the game is analogous to a for-

---

[4] Hilbert did not give an altogether precise account of just what meta-mathematical procedures are to count as finitistic. In the original version of his program the requirements for an absolute proof of consistency were more stringent than in the subsequent explanations of the program by members of his school.

malized mathematical calculus. The pieces and the squares of the board correspond to the elementary signs of the calculus; the legal positions of pieces on the board, to the formulas of the calculus; the initial positions of pieces on the board, to the axioms or initial formulas of the calculus; the subsequent positions of pieces on the board, to formulas derived from the axioms (i.e., to the theorems); and the rules of the game, to the rules of inference (or derivation) for the calculus. The parallelism continues. Although configurations of pieces on the board, like the formulas of the calculus, are "meaningless," statements about these configurations, like meta-mathematical statements about formulas, are quite meaningful. A "meta-chess" statement may assert that there are twenty possible opening moves for White, or that, given a certain configuration of pieces on the board with White to move, Black is mate in three moves. Moreover, general "meta-chess" theorems can be established whose proof involves only a finite number of permissible configurations on the board. The "meta-chess" theorem about the number of possible opening moves for White can be established in this way; and so can the "meta-chess" theorem that if White has only two Knights and the King, and Black has only a King, it is impossible for White to force a mate against Black. These and other "meta-chess" theorems can, in other words, be proved by finitistic methods of reasoning, that is, by examining in turn each of a finite number of configurations that

can occur under stated conditions. The aim of Hilbert's theory of proof, similarly, was to demonstrate by
such finitistic methods the impossibility of deriving certain formally contradictory formulas in a given mathematical calculus.

IV

The Systematic Codification of Formal Logic

There are two more bridges to cross before entering upon Gödel's proof itself. We must indicate how and why Whitehead and Russell's *Principia Mathematica* came into being; and we must give a short illustration of the formalization of a deductive system—we shall take a fragment of *Principia*—and explain how its absolute consistency can be established.

Ordinarily, even when mathematical proofs conform to accepted standards of professional rigor, they suffer from an important omission. They embody principles (or rules) of inference not explicitly formulated, of which mathematicians are frequently unaware. Take Euclid's proof that there is no greatest prime number (a number is prime if it is divisible without remainder by no number other than 1 and the number itself). The argument, cast in the form of a *reductio ad absurdum,* runs as follows:

Suppose, in contradiction to what the proof seeks to establish, that there is a greatest prime number. We designate it by 'x'. Then:

1. x is the greatest prime

2. Form the product of all primes less than or equal to x, and add 1 to the product. This yields a new number y, where $y =$

$$(2 \times 3 \times 5 \times 7 \times \ldots \times x) + 1$$

3. If y is itself a prime, then x is not the greatest prime, for y is obviously greater than x

4. If y is composite (i.e., not a prime), then again x is not the greatest prime. For if y is composite, it must have a prime divisor z; and z must be different from each of the prime numbers 2, 3, 5, 7, ..., x, smaller than or equal to x; hence z must be a prime greater than x

5. But y is either prime or composite

6. Hence x is not the greatest prime

7. There is no greatest prime

We have stated only the main links of the proof. It can be shown, however, that in forging the complete chain a fairly large number of tacitly accepted rules of inference, as well as theorems of logic, are essential. Some of these belong to the most elementary part of formal logic, others to more advanced branches; for example, rules and theorems are incorporated that belong to the "theory of quantification." This deals with relations between statements containing such "quantifying" particles as 'all', 'some', and their synonyms. We shall exhibit one elementary theorem of logic and one rule of inference, each of which is a necessary but silent partner in the demonstration.

Look at line 5 of the proof. Where does it come from? The answer is, from the logical principle (or necessary truth): 'Either p or non-p', where 'p' is called a sentential variable. But how do we get line 5 from this theorem? The answer is, by using the rule of inference known as the "Rule of Substitution for Sentential Variables," according to which a statement can be derived from another containing such variables by substituting any statement (in this case, 'y is prime') for each occurrence of a distinct variable (in this case, the variable 'p'). The use of these rules and logical principles is, as we have said, frequently an all but unconscious action. And the analysis that exposes them, even in such relatively simple proofs as Euclid's, depends upon advances in logical theory made only within the past one hundred years.[5] Like Molière's Monsieur Jourdain, who spoke prose all his life without knowing it, mathematicians have been reasoning for at least two millennia without being aware of all the principles underlying what they were doing. The real nature of the tools of their craft has become evident only within recent times.

For almost two thousand years Aristotle's codification of valid forms of deduction was widely regarded as complete and as incapable of essential improvement. As late as 1787, the German philosopher Immanuel

---

[5] For a more detailed discussion of the rules of inference and logical principles needed for obtaining lines 6 and 7 of the above proof, the reader is referred to the Appendix, no. 2.

Kant was able to say that since Aristotle, formal logic "has not been able to advance a single step, and is to all appearances a closed and completed body of doctrine." The fact is that the traditional logic is seriously incomplete, and even fails to give an account of many principles of inference employed in quite elementary mathematical reasoning.[6] A renaissance of logical studies in modern times began with the publication in 1847 of George Boole's *The Mathematical Analysis of Logic*. The primary concern of Boole and his immediate successors was to develop an algebra of logic which would provide a precise notation for handling more general and more varied types of deduction than were covered by traditional logical principles. Suppose it is found that in a certain school those who graduate with honors are made up exactly of boys majoring in mathematics and girls not majoring in this subject. How is the class of mathematics majors made up, in terms of the other classes of students mentioned? The answer is not readily forthcoming if one uses only the apparatus of traditional logic. But with the help of Boolean algebra it can easily be shown that the class of mathematics majors consists exactly of boys graduating with honors and girls not graduating with honors.

Another line of inquiry, closely related to the work of nineteenth-century mathematicians on the founda-

---

[6] For example, of the principles involved in the inference: 5 is greater than 3; therefore, the square of 5 is greater than the square of 3.

All gentlemen are polite.

No bankers are polite.

No gentlemen are bankers.

$$g \subset p$$
$$b \subset \bar{p}$$
$$\therefore g \subset \bar{b}$$

$$g\bar{p} = 0$$
$$bp = 0$$

$$gb = 0$$

Symbolic logic was invented in the middle of the 19th century by the English mathematician George Boole. In this illustration a syllogism is translated into his notation in two different ways. In the upper group of formulas, the symbol '$\subset$' means "is contained in." Thus '$g \subset p$' says that the class of gentlemen is included in the class of polite persons. In the lower group of formulas two letters together mean the class of things having both characteristics. For example, 'bp' means the class of individuals who are bankers and polite; and the equation '$bp = 0$' says that this class has no members. A line above a letter means "not." ('$\bar{p}$', for example, means "impolite.")

TABLE 1

tions of analysis, became associated with the Boolean program. This new development sought to exhibit pure mathematics as a chapter of formal logic; and it received its classical embodiment in the *Principia Mathematica* of Whitehead and Russell in 1910. Mathemati-

cians of the nineteenth century succeeded in "arithmetizing" algebra and what used to be called the
"infinitesimal calculus" by showing that the various notions employed in mathematical analysis are definable
exclusively in number-theoretical terms (i.e., in terms
of the integers and the arithmetical operations upon
them). For example, instead of accepting the imagi-
nary number $\sqrt{-1}$ as a somewhat mysterious "entity,"
it came to be defined as an ordered pair of integers
(0, 1) upon which certain operations of "addition" and
"multiplication" are performed. Similarly, the irrational number $\sqrt{2}$ was defined as a certain class of ra-
tional numbers—namely, the class of rationals whose
square is less than 2. What Russell (and, before him,
the German mathematician Gottlob Frege) sought to
show was that *all number-theoretical notions* can be defined in purely logical ideas, and that all the axioms of
number theory can be deduced from a small number
of basic propositions certifiable as purely logical truths.

   To illustrate: the notion of *class* belongs to general
logic. Two classes are defined as "similar" if there is a
one-to-one correspondence between their members,
the notion of such a correspondence being explicable
in terms of other logical ideas. A class that has a single
member is said to be a "unit class" (e.g., the class of
satellites of the planet Earth); and the cardinal number
1 can be defined as the class of all classes similar to a
unit class. Analogous definitions can be given of the
other cardinal numbers; and the various arithmetical
operations, such as addition and multiplication, can be

defined in the notions of formal logic. An arithmetical statement, e.g., '1 + 1 = 2', can then be exhibited as a condensed transcription of a statement containing only expressions belonging to general logic; and such purely logical statements can be shown to be deducible from certain logical axioms.

Principia Mathematica thus appeared to advance the final solution of the problem of consistency of mathematical systems, and of number theory in particular, by reducing the problem to that of the consistency of formal logic itself. For, if the axioms of number theory are themselves derivable as theorems in formal logic, the question whether the axioms are consistent is equivalent to the question whether the fundamental axioms of logic are consistent.

The Frege-Russell thesis that mathematics is only a chapter of logic has, for various reasons of detail, not won universal acceptance from mathematicians. Moreover, as we have noted, the antinomies of the Cantorian theory of transfinite numbers can be duplicated within logic itself, unless special precautions are taken to prevent this outcome. But are the measures adopted in *Principia Mathematica* to outflank the antinomies adequate to exclude *all* forms of self-contradictory constructions? This cannot be asserted as a matter of course. Therefore the Frege-Russell reduction of arithmetic to logic does not provide a final answer to the consistency problem; indeed, the problem simply emerges in a more general form. But, irrespective of the validity of the Frege-Russell thesis, two features of

Principia have proved of inestimable value for the further study of the consistency question. *Principia* provides a remarkably comprehensive system of notation, with the help of which all statements of pure mathematics (and of number theory in particular) can be codified in a standard manner; and it makes explicit most of the rules of formal inference used in mathematical demonstrations (eventually, these rules were made more precise and complete). *Principia*, in sum, created the essential instrument for investigating the entire system of number theory as an uninterpreted calculus—that is, as a system of meaningless marks, whose formulas (or "strings") are combined and transformed in accordance with stated rules, of operation.

Because of its historical importance, *Principia Mathematica* will henceforth constitute our prototypical example of a formalization of number theory, and the phrase "und verwandter Systeme" ("and related systems") in the title of Gödel's article will be tacitly included whenever we refer to *Principia Mathematica*: the whole family of such systems will be meant.

V

An Example of a Successful Absolute Proof of Consistency

We must now attempt the second task mentioned at the outset of the preceding section, and familiarize ourselves with an important, though easily understandable, example of an absolute proof of consistency. By mastering the proof, the reader will be in a better position to appreciate the significance of Gödel's paper of 1931.

We shall outline how a small portion of *Principia,* the elementary logic of propositions, can be formalized. This entails the conversion of the fragmentary system into a calculus of uninterpreted signs. We shall then develop an absolute proof of consistency.

The formalization proceeds in four steps. First, a complete catalogue is prepared of the signs to be used in the calculus. These are its vocabulary. Second, the "Formation Rules" are laid down. They declare which of the combinations of the signs in the vocabulary are acceptable as "formulas" (in effect, as sentences). The rules may be viewed as constituting the grammar of the system. Third, the "Transformation Rules" are stated. They describe the precise structure of formulas from which other formulas of given structure are derivable. These rules are, in effect, the rules of inference. Finally, certain formulas are selected as axioms (or as "primitive formulas"). They serve as foundation for the entire system. We shall use the phrase "theorem of the system" to denote any formula that can be derived from the axioms by successively applying the Transformation Rules. By a formal "proof" (or "demonstration") we shall mean a finite sequence of formulas, each of which either is an axiom or can be derived from preceding formulas in the sequence by the Transformation Rules.[7]

For the logic of propositions (often called the "sentential calculus") the vocabulary (or list of "elementary signs") is extremely simple. It consists of variables and constant signs. The variables may have sentences substituted for them and are therefore called "sentential variables." They are the letters

$$\text{'}p\text{'}, \text{'}q\text{'}, \text{'}r\text{'}, \text{etc.}$$

The constant signs are either "sentential connectives" or signs of punctuation. The sentential connectives are:

> '∼' which is short for 'not'
>   (and is called the "tilde"),

---

[7] It immediately follows that axioms are to be counted among the theorems.

'∨'   which is short for 'or',

'⊃'   which is short for 'if . . . then . . . ', and

'·'   which is short for 'and'.

The signs of punctuation are the left- and right-hand round parentheses, '(' and ')', respectively.

The Formation Rules are so designed that combinations of the elementary signs, which would normally have the form of sentences, are called "formulas." Also, each sentential variable counts as a formula. Moreover, if the letter 'S' stands for a formula, its formal negation, namely, $\sim$ (S), is also a formula. Similarly, if S_1 and S_2 are formulas, so are $(S_1) \vee (S_2)$, $(S_1) \supset (S_2)$, and $(S_1) \cdot (S_2)$. Each of the following is a formula: 'p', '$\sim (p)$', '$(p) \supset (q)$', '$((q) \vee (r)) \supset (p)$'. But neither '$(p)(\sim (q))$' nor '$((p) \supset (q)) \vee$' is a formula: not the first, because, while '(p)' and '$(\sim (q))$' are both formulas, no sentential connective occurs between them; and not the second, because the connective '∨' is not, as the Rules require, flanked on both left and right by a formula.[8]

Two Transformation Rules are adopted. One of them, the *Rule of Substitution* (for sentential variables),

---

[8] Where there is no possibility of confusion, punctuation marks (i.e., parentheses) can be dropped. Thus, instead of writing '$\sim (p)$' it is sufficient to write '$\sim p$'; and instead of '$(p) \supset (q)$', simply '$p \supset q$'. (This apparent relaxation of the formality of the system is not really a step away from pure rule-boundness, since the elimination of unneeded parentheses can itself be easily characterized in a purely mechanical manner.)

says that from a formula containing sentential variables it is always permissible to derive another formula
by uniformly substituting formulas for the variables. It
is understood that, when substitutions are made for a
variable in a formula, the same substitution must be
made for *each occurrence* of the variable. For example,
on the assumption that '$p \supset p$' has already been established, we can substitute for the variable 'p' the formula 'q' to get '$q \supset q$'; or we can substitute the formula '$p \vee q$' to get '$(p \vee q) \supset (p \vee q)$'. Or, if we substitute actual English sentences for 'p', we can obtain
each of the following from '$p \supset p$': 'Frogs are noisy $\supset$
Frogs are noisy'; '(Bats are blind $\vee$ Bats eat mice) $\supset$
(Bats are blind $\vee$ Bats eat mice)'.[9] The second Transformation Rule is the *Rule of Detachment* (or *Modus Po-
nens*). This rule says that from two formulas having
the form S_1 and $S_1 \supset S_2$ it is always permissible to derive the formula S_2. For example, from the two formu-
las '$p \vee \sim p$' and '$(p \vee \sim p) \supset (p \supset p)$', we can derive
'$p \supset p$'.

Finally, the axioms of the calculus (essentially those
of *Principia*) are the following four formulas:

----

[9] On the other hand, suppose the formula '$(p \supset q) \supset (\sim q \supset
\sim p)$' has already been established, and we decide to substitute 'r'
for the variable 'p' and '$p \vee r$' for the variable 'q'. We cannot, by
this substitution, obtain the formula '$(r \supset (p \vee r)) \supset (\sim q \supset \sim r)$',
because we have failed to make the same substitution for *each*
occurrence of the variable 'q'. The correct substitution yields
'$(r \supset (p \vee r)) \supset (\sim (p \vee r) \supset \sim r)$'.

1. $(p \lor p) \supset p$
   or, in ordinary English, if either p or p, then p

2. $p \supset (p \lor q)$
   that is, if p, then either p or q

3. $(p \lor q) \supset (q \lor p)$
   that is, if either p or q, then either q or p

4. $(p \supset q) \supset ((r \lor p) \supset (r \lor q))$
   that is, if (if p then q) then (if (either r or p) then (either r or q))

1. If (either Henry VIII was a boor or Henry VIII was a boor) then Henry VIII was a boor

2. If psychoanalysis is fashionable, then (either psychoanalysis is fashionable or headache powders are sold cheap)

3. If (either Immanuel Kant was punctual or Hollywood is sinful), then (either Hollywood is sinful or Immanuel Kant was punctual)

4. If (if ducks waddle then 5 is a prime) then (if (either Churchill drinks brandy or ducks waddle) then (either Churchill drinks brandy or 5 is a prime))

In the left-hand column we have stated the axioms, with a translation for each. In the right-hand column we have given an example for each axiom. The clumsiness of the translations, especially in the case of the final axiom, will perhaps help the reader to realize the advantages of using a special symbolism in formal logic. It is also important to observe that the nonsensical illustrations used as substitution instances for the axioms and the fact that the consequents bear no meaningful relation to the antecedents in the conditional sentences in no way affect the validity of the logical connections asserted in the examples.

Each of these axioms may seem "obvious" and trivial. Nevertheless, it is possible to derive from them with the help of the stated Transformation Rules an indefinitely large class of theorems which are far from obvious or trivial. For example, the formula

'$((p \supset q) \supset ((r \supset s) \supset t)) \supset ((u \supset ((r \supset s) \supset t))$
$$\supset ((p \supset u) \supset (s \supset t)))'$$

can be derived as a theorem. We are, however, not interested for the moment in deriving theorems from the axioms. Our aim is to show that this set of axioms is not contradictory, that is, to prove "absolutely" that it is *impossible* by using the Transformation Rules to derive from the axioms a formula S together with its formal negation $\sim$ S.

Now, it happens that '$p \supset (\sim p \supset q)$' (in words: 'if p, then if not-p then q') is a theorem in the calculus. (We shall accept this as a fact, without exhibiting the derivation.) Suppose, then, that some formula S as well as its formal negation $\sim$ S were deducible from the axioms. By substituting S for the variable 'p' in the theorem (as permitted by the Rule of Substitution), and applying the Rule of Detachment twice, the formula '*q*' would be deducible.[10] But, if the formula consisting of the variable '*q*' is demonstrable, it follows at once that by substituting *any formula whatsoever* for '*q*', *any formula whatsoever is deducible from the axioms.* It is thus clear that, if both some formula S and its formal negation ~ S were deducible from the axioms, every formula would be deducible. In short, if the calculus is not consistent, every formula is a theorem—which is the same as saying that from a contradictory set of axioms any formula can be derived. But this has a converse: namely, if not every formula is a theorem (i.e., if there is at least one formula that is not derivable from the axioms), then the calculus is consistent. *The task, therefore, is to show that there is at least one formula that cannot be derived from the axioms.*

The way this is done is to employ meta-mathematical reasoning upon the system before us. The actual procedure is elegant. It consists in finding a characteristic or structural property of formulas which satisfies the three following conditions. (1) The property must be common to all four axioms. (One such property is that of containing not more than 25 elementary signs; how-

---

[10] By substituting S for '*p*' we first obtain: S ⊃ (~ S ⊃ *q*). From this, together with S, which is assumed to be demonstrable, we obtain by the Detachment Rule: ~ S ⊃ *q*. Finally, since ~ S is also assumed to be demonstrable, using the Detachment Rule once more, we get: *q*.

ever, this property does not satisfy the next condition.)
(2) The property must be "hereditary" under the
Transformation Rules—that is, if all the axioms have
the property, any formula properly derived from them
by the Transformation Rules must also have it. Since
any formula so derived is by definition a theorem, this
condition in essence stipulates that every theorem must
have the property. (3) The property must not belong
to every formula that can be constructed in accordance
with the Formation Rules of the system—that is, we
must seek to exhibit at least one formula that does not
have the property. If we succeed in this threefold task,
we shall have an absolute proof of consistency. The
reasoning runs something like this: the hereditary
property is transmitted from the axioms to all theorems; but if an array of signs can be found that con-
forms to the requirements of being a formula in the
system and that, nevertheless, does not possess the
specified hereditary property, this formula cannot be a
theorem. (To put the matter in another way, if a sus-
pected offspring (formula) lacks an invariably inher-
ited trait of the forebears (axioms), it cannot in fact be
their descendant (theorem).) But, if a formula is discovered that is not a theorem, we have established the
consistency of the system; for, as we noted a moment
ago, if the system were *not* consistent, *every* formula
could be derived from the axioms (i.e., every formula
would be a theorem). In short, the exhibition of a
single formula without the hereditary property does
the trick.

Let us identify a property of the required kind. The one we choose is the property of being a "tautology." In common parlance, an utterance is usually said to be tautologous if it contains a redundancy and says the same thing twice over in different words—e.g., 'John is the father of Charles and Charles is a son of John'. In logic, however, a tautology is defined as a statement that excludes no logical possibilities—e.g., 'Either it is raining or it is not raining'. Another way of putting this is to say that a tautology is "true in all possible worlds." No one will doubt that, irrespective of the actual state of the weather (i.e., regardless of whether the statement that it is raining is true or false), the statement 'Either it is raining or it is not raining' is *necessarily true.*

We employ this notion to define a tautology in our system. Notice, first, that every formula is constructed of elementary constituents 'p', 'q', 'r', etc. A formula is a tautology if it is invariably true, regardless of whether its elementary constituents are true or false. Thus, in the first axiom '$(p \lor p) \supset p$', the only elementary constituent is 'p'; but it makes no difference whether 'p' is assumed to be true or is assumed to be false—the first axiom is true in either case. This may be made more evident if we substitute for 'p' the statement 'Mt. Rainier is 20,000 feet high'; we then obtain as an instance of the first axiom the statement 'If either Mt. Rainier is 20,000 feet high or Mt. Rainier is 20,000 feet high, then Mt. Rainier is 20,000 feet high'. Readers will have no difficulty in recognizing this long statement to be true, even if they should not happen to know whether the constituent statement 'Mt. Rainier is 20,000 feet high' is true. Obviously, then, the first axiom is a tautology—"true in all possible worlds." It can easily be shown that each of the other axioms is also a tautology.

Next, it is possible to prove that the property of being a tautology is hereditary under the Transformation Rules, though we shall not turn aside to give the demonstration. (See Appendix, no. 3.) It follows that every formula properly derived from the axioms (i.e., every theorem) must be a tautology.

It has now been shown that the property of being tautologous satisfies two of the three conditions mentioned earlier, and we are ready for the third step. We must look for a formula that belongs to the system (i.e., is constructed out of the signs mentioned in the vocabulary in accordance with the Formation Rules), yet, because it does not possess the property of being a tautology, cannot be a theorem (i.e., cannot be derived from the axioms). We do not have to look very hard; it is easy to exhibit such a formula. For example, '$p \lor q$' fits the requirements. It purports to be a gosling but is in fact a duckling; it does not belong to the family: it is a *formula,* but it is *not a theorem.* Clearly, it is not a tautology. Any substitution instance (or interpretation) shows this at once. We can obtain by substitution for the variables in '$p \lor q$' the statement 'Napoleon died of cancer or Bismarck enjoyed a cup of coffee'. This is not a truth of logic, because it would be false if both of the two clauses occurring in it were false; and, even if it is a true statement, it is not true irrespective of the truth or falsity of its constituent statements. (See Appendix, no. 3.)

We have achieved our goal. We have found at least one formula that is not a theorem. Such a formula could not occur if the axioms were contradictory. Consequently, it is not possible to derive from the axioms of the sentential calculus both a formula and its negation. In short, we have exhibited an absolute proof of the consistency of the system.[11]

Before leaving the sentential calculus, we must mention a final point. Since every theorem of this calculus is a tautology, a truth of logic, it is natural to ask whether, conversely, every logical truth expressible in the vocabulary of the calculus (i.e., every tautology) is also a theorem (i.e., derivable from the axioms). The answer is yes, though the proof is too long to be stated here. The point we are concerned with making, however, does not

---

[11] The reader may find helpful the following recapitulation of the sequence:

1. Every axiom of the system is a tautology.
2. Tautologousness is a hereditary property.
3. Every formula properly derived from the axioms (i.e., every theorem) is also a tautology.
4. Hence any formula that is not a tautology is not a theorem.
5. One formula has been found (e.g., '$p \lor q$') that is not a tautology.
6. This formula is therefore not a theorem.
7. But, if the axioms were inconsistent, every formula would be a theorem.
8. Therefore the axioms are consistent.

depend on acquaintance with the proof. The point is that, in the light of this conclusion, the axioms are sufficient for generating *all* tautologous formulas— *all* logical truths expressible in the system. An axiomatized system with this property is said to be "complete."

Now, it is frequently of paramount interest to determine whether an axiomatized system is complete. Indeed, a powerful motive for axiomatizing various branches of mathematics has been the desire to establish a set of initial assumptions from which all the true statements in some field of inquiry are deducible. When Euclid axiomatized elementary geometry, he apparently so selected his axioms as to make it possible to derive from them all geometric truths; that is, those that had already been established, as well as any others that might be discovered in the future.[12] Until recently it was taken as a matter of course that a complete set of axioms for any given branch of mathematics can be assembled. In particular, mathematicians believed that the set proposed for number theory in the past was in fact complete, or, at worst, could be made complete simply by adding a finite number of axioms to the original list. The discovery that this will not work is one of Gödel's major achievements.

---

[12] Euclid showed remarkable insight in treating his famous parallel axiom as an assumption logically independent of his other axioms. For, as was subsequently proved, this axiom cannot be derived from his remaining assumptions, so that without it the set of axioms is incomplete.

VI

The Idea of Mapping and Its Use in Mathematics

The sentential calculus is an example of a mathematical system for which the objectives of Hilbert's theory of proof are fully realized. To be sure, this calculus codifies only a fragment of formal logic, and its vocabulary and formal apparatus do not suffice to develop even elementary arithmetic. Hilbert's program, however, is not so limited. It can be carried out successfully for more inclusive systems, which can be shown by meta-mathematical reasoning to be both consistent and complete. By way of example, an absolute proof of consistency is available for a formal system in which axioms for addition but not multiplication are given. But is Hilbert's finitistic method powerful enough to prove the consistency of a system such as *Principia*, whose vocabulary and logical apparatus are adequate to express the whole of number theory and not merely a fragment? Repeated attempts to construct such a proof were unsuccessful; and the publication of Gödel's paper in 1931 showed, finally, that all such efforts operating within the strict limits of Hilbert's original program must fail.

What did Gödel establish, and how did he prove his results? His main conclusions are twofold. In the first place (though this is not the order of Gödel's actual argument), he showed that it is impossible to give a meta-mathematical proof of the consistency of a system comprehensive enough to contain the whole of arithmetic (such as *Principia Mathematica*) —unless the proof itself employs rules of inference different in certain essential respects from the Transformation Rules used in deriving theorems *within* the system. Such a proof may, to be sure, possess great value and importance. However, if the reasoning in it is based on rules of inference much more powerful than the rules of *Principia Mathematica,* so that the consistency of the assumptions in the reasoning is just as subject to doubt as is the consistency of the formalized number theory, the proof would yield only a specious victory: one dragon slain only to create another. In any event, if the proof is not finitistic, it does not realize the aims of Hilbert's original program; and Gödel's argument makes it unlikely that a finitistic proof of the consistency of *Principia Mathematica* (or similar systems) can be given.

Gödel's second main conclusion is even more surprising and revolutionary, because it demonstrates a fundamental limitation in the power of the axiomatic method. Gödel showed that *Principia,* or any other system within which arithmetic can be developed, is *essentially incomplete.* In other words, given *any* consistent formalization of number theory, there are true number-theoretical statements that cannot be derived in the system. This crucial point deserves illustration. Mathematics abounds in general statements to which no exceptions have been found that thus far have thwarted all attempts at proof. A classical illustration is known as "Goldbach's theorem," which states that every even number is the sum of two primes. No even number has ever been found that is not the sum of two primes, yet no one has succeeded in finding a proof that Goldbach's conjecture applies without exception to all even numbers. Here, then, is an example of an arithmetical statement that may be true, but may be non-derivable from the axioms of a formal version of number theory. Suppose, now, that Goldbach's conjecture were indeed universally true, though not derivable from the axioms. What of the suggestion that in this eventuality the axioms or rules of inference could be modified or augmented so as to make hitherto unprovable statements (such as Goldbach's on our supposition) derivable in the enlarged formal system? Gödel's results show that even if the supposition were correct the suggestion would still provide no final cure for the difficulty. That is, even if *Principia Mathematica* were augmented by an indefinite number of new axioms and rules, there will always be further arithmetical truths that are not formally derivable in the augmented system.[13]

_____

[13] Such further truths may, as we shall see, be established by some form of meta-mathematical reasoning about the system. But

How did Gödel prove these conclusions? Up to a point the structure of his argument is modeled, as he himself pointed out, on the reasoning involved in one of the logical antinomies known as the "Richard Paradox," first propounded by the French mathematician Jules Richard in 1905. We shall outline this paradox.

Consider a language (e.g., English) in which the purely arithmetical properties of cardinal numbers can be formulated and defined. Let us examine the definitions that can be stated in the language. It is clear that, on pain of circularity or infinite regress, some terms referring to arithmetical properties cannot be defined explicitly—for we cannot define everything and must start somewhere—though they can, presumably, be understood in some other way. For our purposes it does not matter which are the undefined or "primitive" terms; we may assume, for example, that we understand what is meant by 'an integer is divisible by another', 'an integer is the product of two integers', and so on. The property of being a prime number may then be defined by: 'not divisible by any integer other than 1 and itself'; the property of being a perfect square may be defined by: 'being the product of some integer by itself'; and so on.

We can readily see that each such definition will con-

---

this procedure does not fit the requirement that the calculus must, so to speak, be self-contained, and that the truths in question must be exhibited as the formal consequences of the specified axioms within the system. There is, then, an *inherent* limitation in the axiomatic method as a way of systematizing the whole of number theory.

tain only a finite number of words, and therefore only a finite number of letters of the alphabet. This being the case, the definitions can be placed in serial order: a definition will precede another if the number of letters in the first is smaller than the number of letters in the second; and, if two definitions have the same number of letters, one of them will precede the other on the basis of the alphabetical order of the letters in each. On the basis of this order, a unique integer will correspond to each definition and will represent the number of the place that the definition occupies in the series. For example, the definition with the smallest number of letters will correspond to the number 1, the next definition in the series will correspond to 2, and so on.

Since each definition is associated with a unique integer, it may turn out in certain cases that an integer will possess the very property designated by the definition with which the integer is correlated.[14] Suppose, for instance, the defining expression 'not divisible by any integer other than 1 and itself' happens to be correlated with the order number 17; obviously 17 itself has the property designated by that expression. On the other hand, suppose the defining expression 'being the product of some integer by itself' were correlated

---

[14] This is the same sort of thing as would happen if the English word 'polysyllabic' appeared in a list of words, and we characterized each word of the list by the descriptive tags "monosyllabic" or "polysyllabic". The word 'polysyllabic' would then have the tag "polysyllabic" attached to it.

with the order number 15; 15 clearly does not have the property designated by the expression. We shall describe the state of affairs in the second example by saying that the number 15 has the property of being *Richardian;* and, in the first example, by saying that the number 17 does *not* have the property of being *Richardian.* More generally, we define '*x* is Richardian' as a shorthand way of stating '*x* does *not* have the property designated by the defining expression with which *x* is correlated in the serially ordered set of definitions'.

We come now to a curious but characteristic turn in the statement of the Richard Paradox. The defining expression for the property of being Richardian ostensibly describes a numerical property of integers. The expression itself therefore belongs to the series of definitions proposed above. It follows that the expression is correlated with a position-fixing integer or number. Suppose this number is *n*. We now pose the question, reminiscent of Russell's antinomy: Is *n* Richardian? The reader can doubtless anticipate the fatal contradiction that now threatens. For *n* is Richardian if, and only if, *n* does not have the property designated by the defining expression with which *n* is correlated (i.e., it does not have the property of being Richardian). In short, *n* is Richardian if, and only if, *n* is not Richardian; so that the statement '*n* is Richardian' is both true and false.

We must now point out that the contradiction is, in a sense, a hoax produced by not playing the game quite fairly. An essential but tacit assumption underlying the serial ordering of definitions was conveniently dropped along the way. It was agreed to consider the definitions of the *purely arithmetical* properties of integers—properties that can be formulated with the help of such notions as arithmetical addition, multiplication, and the like. But then, without warning, we were asked to accept a definition in the series that involves reference to the *language* used in formulating arithmetical properties. More specifically, the definition of the property of being Richardian does not belong to the series initially intended, because this definition involves meta-mathematical notions such as the number of letters (or signs) occurring in expressions written in, say, English. We can thus outflank the Richard Paradox by distinguishing carefully between statements *within* arithmetic (which make no reference to any system of notation) and statements *about* some system of notation in which arithmetic is codified.

The reasoning in the construction of the Richard Paradox is clearly fallacious.* The construction nevertheless suggests that it may be possible to "map" or "mirror" meta-mathematical statements about a sufficiently comprehensive formal system in the system it-

---

* Careful consideration of the Richard Paradox shows that it can in fact be reconstructed within the context of a formal system, bypassing the use of a highly ambiguous and ill-defined natural language like English. In such a case, analysis of the fallacy becomes subtler, and it turns out that ideas closely related to those of Gödel's 1931 paper are needed in order to pinpoint the exact step at which the seemingly logical flow of thought goes awry. Such an analysis, however, is beyond the scope of our small book. [—*Ed.*]

self. The idea of "mapping" is well known and plays a fundamental role in many branches of mathematics. It is used, of course, in the construction of ordinary maps, where shapes on the surface of a sphere are projected onto a plane, so that the relations between the plane figures mirror the relations between the figures on the spherical surface. It is used in coordinate geometry, which translates geometry into algebra, so that geometric relations are mapped onto algebraic ones. (The reader will recall the discussion in Section II, which explained how Hilbert used algebra to establish the consistency of his axioms for geometry. What Hilbert did, in effect, was to map geometry onto algebra.) Mapping also plays a role in mathematical physics where, for example, relations between properties of electric currents are represented in the language of hydrodynamics. And mapping is involved when a pilot model is constructed before proceeding with a full-size machine, when a small wing surface is observed for its aerodynamic properties in a wind tunnel, or when a laboratory rig made up of electric circuits is used to study the relations between large-size masses in motion. A striking visual example is presented in Fig. 3, which

Fig. 3

Figure 3 (a) illustrates the Theorem of Pappus: If A, B, C are any three distinct *points* on a *line* I, and A′, B′, C′ any three distinct *points* on another *line* II, the three *points* R, S, T determined by the pairs of *lines* AB′ and A′B, BC′ and B′C, CA′ and C′A, respectively, are *collinear* (i.e., lie on *line* III).

Figure 3 (b) illustrates the "dual" of the above theorem: If A, B, C are any three distinct *lines* on a *point* I, and A′, B′, C′ any three

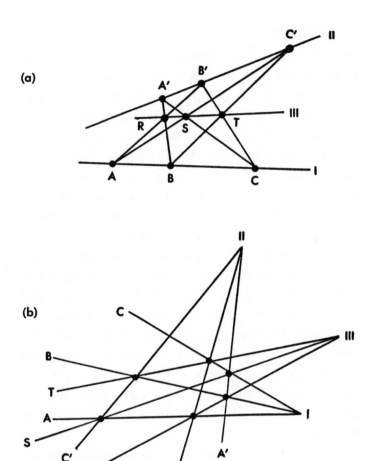

distinct *lines* on another *point* II, the three *lines* R, S, T determined by the pairs of *points* AB′ and A′B, BC′ and B′C, CA′ and C′A, respectively, are *copunctal* (i.e., lie on *point* III).

The two figures have the same *abstract structure,* though in appearance they are markedly different. Figure 3 (a) is so related to Figure 3 (b) that *points* of the former correspond to *lines* of the latter, while *lines* of the former correspond to *points* of the latter. In effect, (b) is a map of (a): a point in (b) represents (or is the "mirror image" of) a line in (a), while a line in (b) represents a point in (a).

illustrates a species of mapping that occurs in a branch of mathematics known as projective geometry.

The basic feature of mapping is that an abstract structure of relations embodied in one domain of "objects" can be shown to hold between "objects" (usually of a sort different from the first set) in another domain. It is this feature which stimulated Gödel in constructing his proofs. If complicated meta-mathematical statements about a formalized system of arithmetic could, as he hoped, be translated into (or mirrored by) arithmetical statements within the system itself, an important gain would be achieved in facilitating meta-mathematical demonstrations. For just as it is easier to deal with the algebraic formulas representing (or mirroring) intricate geometrical relations between curves and surfaces in space than with the geometrical relations themselves, so it is easier to deal with the arithmetical counterparts (or "mirror images") of complex logical relations than with the logical relations themselves.

The exploitation of the notion of mapping is the key to the argument in Gödel's famous paper. Following the style of the Richard Paradox, but carefully avoiding the fallacy involved in its construction, Gödel showed that meta-mathematical statements *about* a formalized arithmetical calculus can indeed be represented by arithmetical formulas *within* the calculus. As we shall explain in greater detail in the next section, he devised a method of representation such that neither the arithmetical formula corresponding to a certain ·meta-mathematical statement, nor the arithmetical formula corresponding to the denial of the statement, is demonstrable within the calculus. Since one of these arithmetical formulas must codify an arithmetical truth, yet neither is derivable from the axioms, the system is incomplete. Gödel's method of representation also enabled him to construct a number-theoretical statement corresponding to the meta-mathematical statement 'The calculus is consistent' and to show that the formal translation of this statement into the notation of the formal calculus is not demonstrable within the calculus. It follows that the meta-mathematical statement cannot be established unless rules of inference are used that cannot be represented within the calculus, so that, in proving the statement, rules must be employed whose own consistency may be as questionable as the consistency of the formal calculus itself. Gödel established these major conclusions by using a remarkably ingenious form of mapping.

VII

Gödel's Proofs

Gödel's paper is difficult. Forty-six preliminary definitions, together with several important preliminary propositions, must be mastered before the main results are reached. We shall take a much easier road; nevertheless, it should afford the reader glimpses of the ascent and of the crowning structure.

A *Gödel numbering*

Gödel described a formalized calculus, which we shall call "PM," within which all the customary arithmetical notations can be expressed and familiar arithmetical relations established.[15] The formulas of the calculus are constructed out of a class of elementary signs, which

---

[15] Gödel used an adaptation of the system developed in *Principia Mathematica.* But any formal system within which the cardinal numbers (i.e., the non-negative whole numbers) and their addition and multiplication can be constructed would have suited his purposes. We shall therefore use the initials "PM" to represent any such system.

constitute the fundamental vocabulary. A set of primitive formulas (or axioms) are the underpinning, and the theorems of the calculus are formulas derivable from the axioms with the help of a carefully enumerated set of Transformation Rules (or rules of inference).

Gödel first showed that it is possible to assign a *unique number* to each elementary sign, each formula (or sequence of signs), and each proof (or finite sequence of formulas). This number, which serves as a distinctive tag or label, is called the "Gödel number" of the sign, formula, or proof.[16]

The elementary signs belonging to the fundamental vocabulary are of two kinds: the constant signs and the variables. We shall assume that there are exactly twelve constant signs,[17] to which the integers from 1 to 12 are attached as Gödel numbers. Most of these signs are already known to the reader: '∼' (short for 'not'); '∨' (short for 'or'); '⊃' (short for 'if . . . then . . . '); '=' (short for 'equals'); '0' (the numeral representing the number zero); '+' (short for 'plus'); '×' (short for 'times'); and three signs of punctuation, namely, the left

---

[16] There are actually many alternative ways of assigning Gödel numbers, and it is immaterial to the main argument which one is adopted. To help the discussion, we give a concrete example of how the numbers can be assigned, and in fact, the method of numbering used in the text is essentially the method employed by Gödel in his 1931 paper.

[17] The number of constant signs depends on how the formal calculus is set up. Gödel in his paper used only seven constant signs. The text uses twelve in order to avoid certain complexities in the exposition. Either way is fine.

parenthesis '(', the right parenthesis ')', and the comma ','. In addition, two other signs will be used: the inverted letter 'Ǝ', which may be read as 'there is', and which occurs in so-called "existential quantifiers"; and the lower-case letter 's', which is prefixed to numerical expressions to designate the immediate successor of a number.

To illustrate: the formula '$(\exists x)\ (x = s0)$' may be read 'There is an x such that x is the immediate successor of zero'. The table below displays the twelve constant signs, states the Gödel number associated with each one, and indicates the usual meaning of the sign.

Constant sign	Gödel number	Usual meaning
~	1	not
∨	2	or
⊃	3	if . . . then . . .
Ǝ	4	there is an . . .
=	5	equals
0	6	zero
s	7	the immediate successor of
(	8	punctuation mark
)	9	punctuation mark
,	10	punctuation mark
+	11	plus
×	12	times

TABLE 2

In light of our statement in Chapter III that symbols in formal calculi are "drained of all meaning" and are merely "empty signs," the reader might well wonder what earthly sense it makes to devote a column to the "meanings" of these supposedly meaningless symbols. Are we not speaking out of both sides of our mouths? The answer is that we are walking a subtle midway path between truly empty signs and truly meaningful ones, which we explain now.

In Table 2, the rightmost column gives each symbol's "usual meaning"—the concept that, through convention, people tend to associate with each symbol. The symbols of PM are, however, fully devoid of meaning in the sense that derivation of theorems depends only on following the formal rules of PM, and never upon taking into account what any of the symbols might stand for. In this sense, PM contains exclusively empty signs. But Russell and Whitehead, given their goal of formalizing mathematics and logic, wanted the symbols of their formal calculus to act as consistently as possible with their conventional interpretations, and so the rules of inference of PM were devised with the goal of making each symbol *merit* its usual, conventional meaning.

To be specific, what makes the meaningless symbol '0' merit the interpretation of "zero," and the meaningless symbol '+' merit the interpretation of "plus"—rather than, say, vice versa? And what would make us feel convinced that the tilde '~', merely a squiggly line that obeys certain formal rules, genuinely represents the abstract concept "not"?

The answer, in a nutshell, is that the interpretation of a symbol hinges on how the symbol behaves inside theorems of PM (and this, in turn, hinges on the axioms and the rules of inference of PM). Thus, for instance, if we can derive theorems such as '0 + 0 = 0', '0 + s0 = s0', and 's0 + s0 = ss0' by following the rules of the formal system, we may start to gain confidence that '0' is acting as one would hope a symbol for zero would act, and that '=' is acting as one would hope a symbol for equality would act, and that '+' is acting as one would hope a symbol for addition would act. Similarly, if the strings '~(0 = s0)', '~~(0 = 0)', and '~(ss0 + ss0 = sss0)' are all theorems of PM, then we will gain some confidence in '~' as a symbol whose natural interpretation is "not." In this manner, theorems collectively pin down their constituent symbols' meanings (or more technically speaking, their symbols' interpretations).

However, having merely a handful of theorems that suggest probable or plausible interpretations for a set of symbols is a far cry from being convinced beyond a shadow of a doubt that these interpretations are absolutely trustworthy. For that, one wants to see large families of truths that are captured by theorems.

In order to lock in the standard interpretations of the symbols in the formal system PM, Gödel showed, in Proposition V of his 1931 paper, that there is an infinite class of theorems of PM, every one of which, if interpreted according to the table of usual meanings above, expresses an arithmetical truth, and conversely, that there is an infinite class of arithmetical truths (the *primitive recursive* ones) every one of which, if it is converted into a formal statement via the table above, yields a theorem of PM.[18] This highly systematic correspondence of truths and interpreted theorems does two things at once: not only does it confirm the power of PM as an axiomatic system for number theory, but it also pins down the conventional interpretations for each and every symbol.

In short, Gödel convincingly demonstrated that the symbols of PM do indeed merit their "meanings" as shown in the third column of Table 2. Today, Gödel's key result is known as the "Correspondence Lemma," the name coming from the two-tier correspondence that it confirms—firstly, that every primitive recursive truth, when encoded as a string of symbols of the formal calculus, is a theorem, and secondly, that on a one-by-one basis, the formal symbols merit their intended interpretations. One sees hereby the way in which truth and meaning are inextricably intertwined.

Besides the constant signs, three kinds of variables appear in PM: the *numerical variables* 'x', 'y', 'z', etc., for which numerals (such as '$ss0$') and numerical expres-

---

[18] The infinite class of primitive recursive truths includes all correct additions, all correct multiplications, and a vast variety of statements such as "17 is the 7th prime," "21 is not a prime number," and so forth. The fact that all primitive recursive truths yield theorems of PM guarantees that the meanings we have assigned to the symbols of PM are merited.

sions (such as '$x + y$') may be substituted; the *sentential variables* 'p', 'q', 'r', etc., for which formulas (sentences) may be substituted; and the *predicate variables* 'P', 'Q', 'R', etc., for which predicates, such as "is prime", or "is greater than," may be substituted. The variables are assigned Gödel numbers in accordance with the following rules: (i) with each distinct numerical variable, associate a distinct prime number greater than 12; (ii) with each distinct sentential variable, associate the square of a prime number greater than 12; (iii) with each distinct predicate variable, associate the cube of a prime number greater than 12. The table below illustrates these rules.

Numerical variable	Gödel number	A possible substitution instance
x	13	0
y	17	s0
z	19	y

Numerical variables are associated with prime numbers greater than 12.

Sentential variable	Gödel number	A possible substitution instance
p	13^2	$0 = 0$
q	17^2	$(\exists x)\ (x = sy)$
r	19^2	$p \supset q$

Sentential variables are associated with squares of primes greater than 12.

Predicate variable	Gödel number	A possible substitution instance
P	13^3	$x = sy$
Q	17^3	$\sim(x = ss0 \times y)$
R	19^3	$(\exists z)\ (x = y + sz)$

Predicate variables are associated with cubes of primes greater than 12.

TABLE 3

Consider next a formula belonging to PM—for example, '$(\exists x)\ (x = sy)$'. (Literally translated, this reads: 'There is an x such that x is the immediate successor of y', and it says, in effect, that whatever number the variable y happens to stand for has an immediate successor.) The numbers associated with its ten constituent elementary signs are, respectively: 8, 4, 13, 9, 8, 13, 5, 7, 17, 9. We show this schematically below:

(	∃	x	)	(	x	=	s	y	)
↓	↓	↓	↓	↓	↓	↓	↓	↓	↓
8	4	13	9	8	13	5	7	17	9

It is of the essence, however, to assign a *single* number to the formula rather than a sequence of numbers. Fortunately, this can be done easily. We agree to associate with the formula the unique number that is the product of the first ten primes in order of magnitude, each prime being raised to a power equal to the Gödel number of the corresponding sign. The formula is accordingly associated with the number

$$2^8 \times 3^4 \times 5^{13} \times 7^9 \times 11^8 \times$$
$$13^{13} \times 17^5 \times 19^7 \times 23^{17} \times 29^{\,9}.$$

Let us refer to this very large number as m. In a similar fashion, a unique Gödel number—the product of as many successive primes as there are signs, each prime being raised to a power equal to the Gödel number of the corresponding sign—can be assigned to every finite sequence of elementary signs, and, in particular, to every formula.[19]

Consider, finally, a *sequence* of formulas, such as may occur in some proof—e.g., the sequence:

$$(\exists x)\ (x = sy)$$
$$(\exists x)\ (x = s0)$$

---

[19] Signs may occur in PM which do not appear in the fundamental vocabulary; these are introduced by defining them with the help of the elementary signs. For example, the sign '·', the sentential connective used as an abbreviation for 'and', can be defined as follows: '$p \cdot q$' is short for '$\sim(\sim p \lor \sim q)$'. What Gödel number is assigned to such a defined sign? The answer is obvious if we notice that expressions containing defined signs can be eliminated in favor of their defining equivalents; and it is clear that a Gödel number can be determined for the transformed expressions. Accordingly, the Gödel number of the formula '$p \cdot q$' will simply be the Gödel number of the formula '$\sim(\sim p \lor \sim q)$'. Similarly, the numerals of the decimal system can be introduced by definition as follows: '1' as short for 's0', '2' as short for 'ss0', '3' as short for 'sss0', and so on. To obtain the Gödel number for the formula '$\sim(2 = 3)$', we eliminate the defined signs, thus obtaining the pure PM formula '$\sim(ss0 = sss0)$', whereupon we determine its Gödel number pursuant to the rules stated in the text.

The lower of these formulas, when translated, reads: 'Zero has an immediate successor'; it is mechanically derivable from the upper formula via one of the rules of inference of PM, which states that it is valid to substitute any numerical expression (here, the numeral '0') for a numerical variable (here, the variable 'y').[20]

We have already determined the Gödel number of the upper formula: it is m; let us now suppose that n is the Gödel number of the lower formula. As before, it is of the essence to have a single number, rather than a sequence of numbers, as a tag for any given sequence of formulas. We agree therefore to associate with this particular sequence of formulas the number which is the product of the first two primes in order of magnitude (i.e., the primes 2 and 3), each prime being raised to a power equal to the Gödel number of the corresponding number in the sequence. Thus, if we call this number k, we can write $k = 2^m \times 3^n$. By applying this simple procedure, we can obtain a number for any sequence of formulas. In sum, every expression in

---

[20] The reader will recall that we defined a proof as a finite sequence of formulas, each of which either is an axiom or can be derived from preceding formulas in the sequence with the help of the Transformation Rules of PM. By this definition, the above sequence is not a proof, since the first formula is not an axiom and its derivation from the axioms is not shown: the sequence is only a short segment of a proof. It would take too long to write out a full example of a proof, and for illustrative purposes the above example will suffice.

the formal system PM—whether it is an elementary sign, a sequence of signs, or a sequence of such sequences—can be assigned a unique Gödel number.

What has been done so far is to establish a method for completely "arithmetizing" the formal calculus. The method is essentially a set of directions for setting up a one-to-one correspondence between the expressions in the calculus and a certain subset of the positive integers.[21] Once an expression is given, the Gödel number uniquely corresponding to it can be calculated.

But this is only half the story. Once a number is given, we can determine whether it is a Gödel number, and if it is, the precise expression that it represents can be "retrieved" from it. If a given number is less than or equal to 12, it is the Gödel number of an elementary constant sign. The sign can be identified using Table 2. If the number is greater than 12, it can be decom-

---

[21] Not every positive integer is a Gödel number. Consider, for example, the number 100. Since 100 is greater than 12, it cannot be the Gödel number of an elementary constant sign, and since it is neither a prime greater than 12, nor the square nor the cube of such a prime, it cannot be the Gödel number of a variable of any sort. On decomposing 100 into its prime factors, we find that it is equal to $2^2 \times 5^2$, and we see that the prime number 3 does not appear as a factor in the decomposition, but is skipped. According to the rules laid down, however, the Gödel number of a formula (or of a sequence of formulas) must be the product of *successive* primes, each raised to some power. The number 100 does not satisfy this condition, however. In short, 100 does not map, via the rules of Gödel numbering, onto any constant sign, variable sign, or formula, and hence 100 is not a Gödel number.

posed into its prime factors in just one way (as we know from a famous result of number theory).[22] If it is a prime greater than 12, or the second or third power of such a prime, then it is the Gödel number of an identifiable variable. And finally, if it is the product of successive primes, each raised to some power, it may be the Gödel number either of a formula or of a sequence of formulas. In this case the expression to which it corresponds can be exactly determined.

A	243,000,000
B	$64 \times 243 \times 15,625$
C	$2^6 \times 3^5 \times 5^6$
D	$\begin{matrix} 6 & 5 & 6 \\ \downarrow & \downarrow & \downarrow \\ 0 & = & 0 \end{matrix}$
E	$0 = 0$

The formula of PM that expresses the idea "zero equals zero" has the Gödel number 243 million. When read downwards from A to E, the illustration shows how a number can be translated into the expression for which it stands; when read upwards, it shows how to calculate the number that stands for a given formula.

TABLE 4

----

[22] This result is known as the *fundamental theorem of arithmetic*, and it states that if an integer is composite (i.e., not a prime), it has a unique decomposition into prime factors, with associated exponents.

Following this program, we can take any given number apart, as if it were a machine, discover how it is constructed and what goes into it; and since each of its elements corresponds to an element of the expression it represents, we can reconstitute the expression, analyze its structure, and the like. Table 4 illustrates for a sample positive integer how we can ascertain whether it is a Gödel number and if so, what expression it symbolizes.

B *The arithmetization of meta-mathematics*

Gödel's next step is an ingenious application of mapping. He showed that all meta-mathematical statements about the structural properties of expressions in the formal calculus can be accurately *mirrored* within the calculus itself. The basic idea underlying his procedure is this: Since every expression in PM is associated with a particular (Gödel) number, a meta-mathematical statement about formal expressions and their *typographical* relations to one another may be construed as a statement about the corresponding (Gödel) numbers and their *arithmetical* relations to one another. In this way meta-mathematics becomes completely "arithmetized." To take a trivial analogue: customers in a busy supermarket are often given, when they enter, tickets on which are printed numbers whose order determines the order in which the customers are to be waited on at the meat counter. By inspecting the numbers it is easy to tell how many persons have been served, how many are waiting, who precedes whom, and by how many customers, and so on. If, for example, Mr. Nagel has number 37 and Mr. Newman has number 53, instead of explaining to Mr. Newman that he has to wait his turn after Mr. Nagel, it suffices to point out that 37 is less than 53.

As in the supermarket, so in meta-mathematics. More concretely, the exploration of meta-mathematical questions can be pursued alternatively (or indirectly) by investigating certain arithmetical properties and relations of (rather large) whole numbers. We shall see that each meta-mathematical statement about strings of symbols and how they are *typographically* related (e.g., a sequence of three particular formulas constitutes a proof of a fourth particular formula) corresponds to a statement about the strings' Gödel numbers and how those numbers are *arithmetically* related.

We illustrate these general remarks by a very elementary example. Consider the simple formula '$\sim(0 = 0)$', expressing the blatant falsity that zero does not equal itself. We now make the true meta-mathematical observation that the first symbol of this formula is the tilde '$\sim$'. If, thanks to Gödel numbering, meta-mathematics can indeed be faithfully mapped into the domain of integers and their properties, then surely this true observation must map onto a true number-theoretical assertion. The question is, which one? To find the answer, we first need the Gödel number of the formula in question—namely, $2^1 \times 3^8 \times 5^6 \times 7^5 \times 11^6$

$\times$ 13⁹, which we shall call '*a*'. Clearly, the statement we are seeking has to do with the decomposition of this huge number into its prime factors—in particular, it is the assertion that the exponent of the smallest prime (namely, 2) in *a*'s prime factorization is 1. In other words, the desired assertion about numbers is this: '2 is a factor of *a,* but 2^2 is not a factor of *a.*'

We have found the number-theoretical way of saying that our formula's first symbol is the tilde. That is a key step, but the next step is equally important, and it consists in converting this informal statement of English into a formal string of PM. How does one express the predicate '*x* is a factor of *y*' in PM notation? That, fortunately, is very easy: it is accomplished by restating the target as 'There is a number *z* such that *y* equals *z* times *x*'—which carries over directly into the PM formula '$(\exists z)\ (y = z \times x)$'. In our case, we need to make use of this predicate twice, once prefixed by a tilde:

$(\exists z)\ (sss \ldots sss0 = z \times ss0)\ \cdot$
$$\sim(\exists z)\ (sss \ldots sss0 = z \times (ss0 \times ss0))$$

The long numeral that here occurs twice must of course contain exactly *a* copies of 's'. Note the dot in the middle, symbolizing the concept 'and' (see footnote 19). Thus our formula says, quite literally, 'There is a number *z* such that *a* equals *z* times 2, and there is no number *z* such that *a* equals *z* times the quantity 2×2'.

Although a bit ponderous, this formula constitutes

PM's way of expressing a simple meta-mathematical assertion about the identity of the first symbol of another one of its own formulas. Never mind that the above formula, if it were written out in full, would be unimaginably huge; despite its astronomic proportions, it is conceptually a very simple formula. Furthermore, since the arithmetical predicate 'x is a factor of y' is primitive recursive (the reader must take this on faith), the Correspondence Lemma guarantees that the string above, expressing a truth about numbers, is a *theorem* of PM.

In sum, there is a theorem of PM that is a translation of the true meta-mathematical statement "The initial symbol of '$\sim (0 = 0)$' is the tilde." We thus see our first example of how PM is actually able to talk about itself faithfully (i..e, with meta-mathematical truths being mirrored by theorems of PM), thanks firstly to the ingenious mapping device of Gödel numbering, and secondly to the Correspondence Lemma, guaranteeing that symbols merit their meanings.

The extremely simple case we have just given exemplifies a very general and deep insight that lies at the heart of Gödel's discovery: typographical properties of long chains of symbols can be talked about in an indirect but perfectly accurate manner by instead talking about the properties of prime factorizations of large integers. This is what we mean by the phrase "the arithmetization of meta-mathematics." When we combine this idea with that of the formalization of arithmetic (i.e., number theory) within PM, we arrive at the idea of the formalization of meta-mathematics within PM.

Let us now turn our attention to a more complex type of meta-mathematical statement: 'The sequence of formulas with Gödel number x is a proof (in PM) of the formula with Gödel number z'. Thanks to the arithmetization of meta-mathematics, this statement about typographical relations between certain strings is mirrored inside number theory by a statement about purely numerical relationships between the two numbers x and z. (We can gain gain some notion of the complexity of this numerical relationship by recalling the example above, in which the Gödel number $k = 2^m \times 3^n$ was assigned to the (fragment of a) proof whose conclusion—i.e., last line—has the Gödel number n. A little reflection shows that there is here a definite, though by no means simple, arithmetical relation between k, the Gödel number of the proof, and n, the Gödel number of the conclusion.) We shall denote this arithmetical relationship between the integers x and z by the abbreviation 'dem (x, z)'. The lowercase letters 'dem' were chosen to remind us of the meta-mathematical relationship to which this number-theoretical relationship corresponds—namely, 'The sequence of formulas with Gödel number x is a proof (i.e., a *demonstration*) inside PM of the formula with Gödel number z'.

Note that the numerical relationship denoted by 'dem' depends implicitly on all the axioms and rules of inference of PM. If we were to modify PM in some manner, then the notion of "proof" would be slightly different, and accordingly it would map onto a slightly different numerical relationship, but still very similar to dem, which would play the same role for the changed system as dem plays for PM.

Gödel took great pains, in his paper, to convince his readers that dem (x, z) is a primitive recursive relationship between the numbers x and z, and from this fact (which we shall accept on faith) it follows, via Gödel's Correspondence Lemma, that there is a formula of PM expressing this relationship in the formal notation. We shall denote this formula by the abbreviation 'Dem (x, z)', with a capital 'D', which signals its formality.

Note carefully that whereas 'dem $(2, 5)$' is a meaningful statement about the integers 2 and 5 (meaningful but patently false, since 2 is not the Gödel number of any proof, and 5 is not the Gödel number of a full formula), its formal counterpart—'Dem (ss0, sssss0)'— is a mere string of PM, and hence, strictly speaking, is neither true nor false but simply meaningless.[23] Once

---

[23] For a simpler example of this crucial distinction between the formal and informal levels involved in our discussion, consider the arithmetical assertion "two plus two does not equal five." This is not a string of PM but a statement in English, and it happens to be true. It can be written more concisely as '$2 + 2 \neq 5$', but this is still informal and its constituent symbols are taken as having meanings. There is a formal counterpart, however—'$\sim$(ss0 + ss0 = sssss0)'— which, strictly speaking, is just a string of empty signs and is therefore neither true nor false, but meaningless. On the other hand, addition being primitive recursive, the Correspondence

again, Gödel's Correspondence Lemma enters the picture and assures us that for any true instance of the number-theoretical predicate dem (x, z) there is a theorem of the form 'Dem (sss . . . sss0, sss . . . sss0)', where the first string of 's''s is of length x, and the second string of 's''s is of length z.

The existence of the formula 'Dem (x, z)' inside PM tells us something very crucial: that true meta-mathematical assertions of the form "such-and-such demonstrates so-and-so by the rules of PM" are faithfully reflected in theorems of PM. By the same token,

---

Lemma assures us that this formula is a theorem of PM. We sometimes loosely say of a PM formula such as this one that it is true (or false), meaning that the arithmetical statement that it expresses is true (or false). In that loose parlance, '$\sim$(ss0 + ss0 = sssss0)' would be true.

A more complex example involves the concept of primeness. Let us denote the number-theoretical predicate 'x is prime' by 'pr (x)'. The (false) statement claiming "nine is prime" would then be 'pr (9)'. This is not a string of PM but merely a convenient shortening of the English sentence. There is, however, a formal counterpart to 'pr (x)' inside PM, and we can even exhibit it: '$\sim (\exists y)\ (\exists z)$ $(x = ssy \times ssz)$'. (To be sure, this is but one of many possible ways of expressing primeness inside PM; readers should make sure they understand why this particular formula works.) We can denote this formula by 'Pr (x)', with a capital 'P'. Then the false claim that nine is prime would be expressed inside PM as 'Pr (sssssssss0)'. Note that the full formula for which this stands, if written out, would be '$\sim(\exists y)\ (\exists z)$ (sssssssss0 $= ssy \times ssz$)'. Since primeness is primitive recursive, the Correspondence Lemma assures us that the string '$\sim$Pr (sssssssss0)'—which is merely a shorthand for '$\sim\sim (\exists y)\ (\exists z)$ (sssssssss0 $= ssy \times ssz$)'—is a theorem of PM.

every true meta-mathematical statement of the form "such-and-such does *not* demonstrate so-and-so by the rules of PM" is faithfully mirrored by a PM theorem of the form '~Dem (sss . . . sss0, sss . . . sss0)', with, as usual, the appropriate numbers of 's''s. Once again, then, thanks to Gödel's mapping, PM is seen to have the capability of talking accurately about itself.

One final special concept and corresponding piece of notation will be needed before we can state the crux of Gödel's argument. Let us begin with an example. The formula '$(\exists x)$ $(x = sy)$' has the very large number m for its Gödel number, as we saw a few pages back, and one of the variables inside it has the Gödel number 17. Suppose we were to replace, in our formula, the variable with Gödel number 17 (i.e., the variable 'y') by the numeral for m itself. The result would be the extremely long formula '$(\exists x)$ $(x = sss . . . sss0)$', where the string of 's''s consists of $m + 1$ copies. (Translated into English, this new string of PM asserts that there is a number x such that x is the immediate successor of m—or in short, that m has a successor.)

This very long formula itself has a Gödel number, and of course that number is very large, but no matter how large it is, it can, in principle, be calculated perfectly straightforwardly. But instead of worrying about the details of the calculation or its exact result, we can simply characterize the resultant number in an unambiguous meta-mathematical fashion: it is the Gödel number of the formula that is obtained from the formula with Gödel number m, by substituting for the variable with Gödel number 17 the numeral for m itself. This characterization uniquely determines a specific positive integer as a function of the numbers m and 17.[24]

---

[24] This function is quite complex. Just how complex becomes evident if we try to formulate it in greater detail. Let us attempt such a formulation, without carrying it to the bitter end. We saw earlier that m, the Gödel number of '$(\exists x)(x = sy)$', is

$$2^8 \times 3^4 \times 5^{13} \times 7^9 \times 11^8 \times 13^{13} \times 17^5 \times 19^7 \times 23^{17} \times 29^9.$$

To find the Gödel number of the modified formula, with the numeral for m substituted for the variable 'y', we must look at the signs in that formula one by one and raise successive prime numbers to the corresponding powers. Recall that the formula of interest is: '$(\exists x)(x = sss \ldots sss0)$', with $m + 1$ copies of 's'. The Gödel numbers of the individual symbols are thus:

$$8, 4, 13, 9, 8, 13, 5, 7, 7, 7, 7, 7, \ldots, 7, 7, 7, 7, 7, 6, 9.$$

In this sequence, the number 7 occurs $m + 1$ times. We now raise the appropriate primes to these powers, one by one:

$$2^8 \times 3^4 \times 5^{13} \times 7^9 \times 11^8 \times 13^{13} \times 17^5 \times$$
$$19^7 \times 23^7 \times 29^7 \times 31^7 \times \ldots \times (p_{m+10})^9$$

(Here, p_{m+10} is the $(m + 10)$th prime in order of magnitude.)

Let us give this very enormous number the name 'r'. Now compare the two Gödel numbers m and r. The former contains a prime factor raised to the factor 17 (because the initial formula contained the variable 'y'), while the latter contains all the prime factors of m and many others besides, but none of them are raised to the 17th power. The number r can thus be obtained from the number m by replacing the prime factor in m that is raised to the 17th power by other primes raised to various powers different from 17. To state exactly and in full detail just how r is related to m is not possible without introducing a great deal of additional notational apparatus; this is done in Gödel's paper. But hopefully enough has been said

This rather circular-seeming idea of substituting a string's own Gödel number into the string itself (and then taking the Gödel number of the result) was one of Gödel's key insights, as we shall see, and he again took great pains to convince his readers that this function is sufficiently straightforwardly calculable as to be primitive recursive, and thus to fall under the scope of the Correspondence Lemma. We will designate the new Gödel number as a function of the old Gödel number x by the notation 'sub $(x, 17, x)$'. Although to do so involves a mouthful, we can pinpoint exactly what this number is: it is the Gödel number of the formula obtained by taking the formula with Gödel number x and, wherever there are occurrences of the variable 'y' in that formula, replacing them by the numeral for x.[25]

---

to convince the reader that the number r is a well-defined number-theoretical function of m and 17.

[25] It may be asked why, in the meta-mathematical characterization just mentioned, we say that it is "the *numeral* for x" which is to be substituted for a certain variable, rather than "the *number* x". The answer depends on the distinction, already discussed, between mathematics and meta-mathematics, and calls for a brief elucidation of the difference between numbers and numerals. A *numeral* is a *sign*, a linguistic expression, something which one can write down, erase, copy, and so on. A *number,* on the other hand, is something which a numeral *names* or *designates,* and which cannot literally be written down, erased, copied, and so on. Thus, we say that ten is the *number* of our fingers, and, in making this statement, we are attributing a certain "property" to the class of our fingers; but it would clearly be absurd to say that this property is a numeral.

Given that Gödel's sub function is a primitive recursive one, there exists some formal expression inside PM that mirrors it exactly,[26] and we shall abbreviate that expression by 'Sub $(x, 17, x)$', drawing the crucial distinction between the informal arithmetical concept and its formal typographical counterpart as we did before—namely, by using lowercase and uppercase initial letters, respectively. You should keep in mind that whereas 'sub $(243,000,000, 17, 243,000,000)$' designates a *number* (i.e., a magnitude or quantity),[27] what

---

Again, the number ten is named by the Arabic numeral '10', as well as by the Roman letter 'X'; these names are different, though they name the same number. In short, when we make a substitution for a numerical variable (which is a letter or sign), we are putting one sign in place of another sign. We cannot literally substitute a number for a sign, because a number is a *concept* (and is sometimes said to be an abstract property of classes), rather than something we can put on paper. It follows that, in substituting for a numerical variable in a formula, we can insert only a numeral (or some other numerical expression, such as '0 $\times$ 0' or 'ss0 + sss0'), and not a number. This explains why, in the meta-mathematical characterization of the sub function, it was stated that we are to replace occurrences of the variable 'y' by the *numeral* for (the number) x, rather than by the *number x* itself. Despite striving for this kind of linguistic precision, we can easily slip into speaking of substituting a number for a variable in a formula; sometimes it is actually clearer to speak in this loose manner.

[26] Strictly speaking, what the Correspondence Lemma applies to is not the function sub, but the predicate '$z = $ sub $(x, 17, x)$'; this distinction, however, is such a minor detail as to merit no more than a footnote. Incidentally, this detail is related to the point made in footnote 28.

the abbreviation 'Sub (243,000,000, 17, 243,000,000)' designates is a *string* inside PM. And although that string is, strictly speaking, meaningless (as, of course, are all strings in PM or in any other formal system), it is convenient to think of it as having a meaning, because it acts as the formal representative for a certain fairly involved arithmetical calculation, in much the same way as the "meaningless" string 'ss0 + ss0' acts as the formal representative inside PM for the very simple calculation "two plus two" (and thus, although more indirectly, for the concept "four").[28]

---

[27] Readers may wonder what number is designated by 'sub (x, 17, x)' if the formula whose Gödel number is x happens not to contain the variable with Gödel number 17—that is, if the initial formula does not contain the variable 'y' anywhere. Thus, sub (243,000,000, 17, 243,000,000) is the Gödel number of the formula made from the formula with Gödel number 243,000,000 by substituting for the variable 'y' the numeral 'sss . . . sss0' (containing 243,000,000 copies of 's'). But if you consult Table 4, you will see that 243,000,000 is the Gödel number of the string '0 = 0', which has no 'y'. What, then, is the formula obtained from '0 = 0' by substituting for 'y' the given numeral? The answer is simple: since the formula '0 = 0' has no 'y', no substitution can be made, and hence the "modified" formula is just the formula itself, untouched. Accordingly, the number designated by 'sub (243,000,000, 17, 243,000,000)' is just 243,000,000.

[28] The reader may be puzzled as to whether 'Sub (x, 17, x)' is a *formula* of PM, in the sense that, for example, 's0 = sss0', '($\exists x$)(x = sy)', and 'Dem (x, z)' are formulas. The answer is no, for the following reason. The string 's0 = sss0' is a formula because it

C *The heart of Gödel's argument*

At last we are equipped to follow in outline Gödel's main argument. We shall begin by enumerating the steps in a general way, so that the reader can get a bird's-eye view of the sequence.

Gödel showed (i) how to construct a formula G of PM that represents the meta-mathematical statement: 'The formula G is not demonstrable using the rules of PM'.[29] This formula thus ostensibly says *of itself* that it is not demonstrable. Up to a point, G is constructed analogously to the Richard Paradox. In that paradox, the expression 'Richardian' is associated with a certain number n, and the sentence 'n is Richardian' is constructed. In Gödel's argument, the formula G is likewise associated with a certain number g—namely, its Gödel number—and G is so constructed that it says

---

asserts a relation between two numbers and is thus capable of having truth or falsity attributed to it. Similarly, when numerals are substituted for the variables in the string 'Dem (x, z)', the resulting formula expresses an arithmetical statement about two numbers, and this statement is either true or false. Much the same holds for '$(\exists x)\ (x = sy)$'. On the other hand, when a numeral is substituted for 'x' in the string 'Sub $(x, 17, x)$', the resulting string does not *assert* anything and therefore cannot be assigned a truth value. For this reason, 'Sub $(x, 17, x)$' is not a formula. Like the string 'ss0 $\times$ sssss0', it merely *designates* or *names* a number, by describing it as a certain *function* of other numbers.

[29] From now on, whenever we write "demonstrable" without any further qualifier, it should always be taken as meaning "demonstrable using the rules of PM" (and is synonymous with "is a theorem of PM").

'The formula that has Gödel number *g* is not demonstrable'.

But (ii) Gödel also showed that G is demonstrable if, and only if, its formal negation ~G is demonstrable. This step in the argument is again analogous to the step in the Richard Paradox in which it is proved that *n* is Richardian if, and only if, *n* is not Richardian. However, if a formula and its own negation are both formally demonstrable, then PM is not consistent. Accordingly, if PM is consistent, neither G nor ~G can be formally derivable from the axioms. In short, if PM is consistent, then G is a *formally undecidable* formula.[30]

Gödel then showed (iii) that, though G is not formally demonstrable, it nevertheless is a *true* arithmetical formula (see the remarks about loose parlance in footnote 23). G is true in the sense that it claims that a certain arithmetical property defined by Gödel is possessed by no integer—and indeed, no integer possesses the property, as Gödel shows.

Step (iv) is the realization that since G is both true and formally undecidable (within PM), PM must be *incomplete.* In other words, we cannot deduce all arithmetical truths from the axioms and rules of PM. Moreover, Gödel established that PM is *essentially* incom-

---

[30] To assert that some formula X is "formally undecidable," as in the title of Gödel's paper (or simply "undecidable" for short), means that neither X nor its negation ~X is demonstrable inside the formal calculus of interest—e.g., "PM or related systems" (as in the article's title).

plete: even if PM were augmented by additional axioms (or rules) so that the true formula G could be formally derived within the enhanced calculus, then another true formula G′ could be constructed in a precisely analogous manner, and G′ would be formally undecidable inside the enhanced calculus. Needless to say, further enhancement of the already-enhanced calculus, so as to allow derivation of G′, would merely lead to yet another formula G″ undecidable within the doubly augmented system—and so on, ad infinitum. This is the meaning of "essentially incomplete."

In step (v), Gödel described how to construct a formula A of PM that represents the meta-mathematical statement: 'PM is consistent'; and he showed that the formula 'A ⊃ G' is formally demonstrable inside PM. Finally, he showed that the formula A is not demonstrable inside PM. From this it follows that the consistency of PM cannot be established by any chain of logical reasoning that can be mirrored within the formal reasoning system that PM itself constitutes.

It is worth adding that Gödel was crucially concerned with the generality of his results, which is why in the title of his paper he explicitly stated that his results pertained not only to Russell and Whitehead's celebrated formal system but also to "related systems." He wrote at the end of his paper, "Throughout this work we have virtually confined ourselves to the system PM, and have merely indicated the applications to other systems. The results will be stated and proved in fuller generality in a forthcoming sequel." Gödel was in fact worried that because of the shock value of his paper, many people would doubt its validity, and he therefore intended to buttress his reasoning in the sequel. It turned out, however, that his paper was so cogently written that its conclusions were quickly accepted, thus obviating any need for a sequel. The point, then, is that Gödel's results are not due to some odd defect in the specific system PM; they are applicable to *any* system that incorporates the arithmetical properties of the cardinal numbers, including addition and multiplication.

Now we shall broach Gödel's argument more fully.

(i) The formula 'Dem (x, z)' has already been defined. It represents within PM the meta-mathematical statement: 'The sequence of formulas with the Gödel number x is a proof for the formula with the Gödel number z'. Let us now prefix this formula with an existential quantifier, as follows: '$(\exists x)$ Dem (x, z)'. The interpretation of this formula is straightforward: 'There exists a sequence of formulas (with Gödel number x) that constitutes a demonstration of the formula with Gödel number z'. More compactly: 'The formula with Gödel number z is demonstrable'. (We remind readers that in this context, the terms 'demonstration' and 'demonstrable' always refer to the formal system PM.)

If we prefix this formula with the tilde, thus constructing its formal negation, we get: '$\sim(\exists x)$ Dem (x, z)'. This formula constitutes a formal paraphrase, within PM, of the meta-mathematical statement: 'The formula with Gödel number z is not demonstrable' — or, to put it another way, 'No proof can be adduced for the formula with Gödel number z'.

What Gödel showed is that a certain special case of this formula is not formally demonstrable. To construct this special case, we begin with the formula displayed as line (1):

(1) $\sim(\exists x)$ Dem $(x, \text{Sub } (y, 17, y))$

This formula belongs to PM, but it possesses a metamathematical interpretation. The question is, which one? The reader should recall that the expression 'Sub $(y, 17, y)$' designates a number. This number is the Gödel number of the formula obtained from the formula with Gödel number y by substituting for the variable with Gödel number 17 (i.e., for all occurrences of the letter 'y') the numeral for y.[31] It will then be evident that the formula on line (1) represents the metamathematical statement: 'The formula with Gödel number sub $(y, 17, y)$ is not demonstrable'. Although this is a tantalizing statement, it is still open-ended and indefinite, since it still contains the variable 'y'. To

___

[31] It is crucial to recognize that 'Sub $(y, 17, y)$', though an expression of PM, is not a formula but a name-function for identifying a *number* (see footnote 28). The number so identified will be the Gödel number of a specific formula. Or rather, it would be, were 'y' not a variable. Since 'y' is a variable and not a numeral, the expression 'Sub $(y, 17, y)$' doesn't represent a specific number any more than does the string '$y + sss0$'. For that, the variable 'y' would need to be replaced by a specific numeral.

make it definite, we need a numeral in place of a variable. What numeral should we choose? Here we shall follow Gödel.

Since the formula on line (1) belongs to PM, it has a (very large) Gödel number that could, in principle, be calculated. Luckily, we shall not actually calculate it (nor did Gödel); we shall simply designate its value by the letter 'n'. We now proceed to replace all occurrences of the variable 'y' in formula (1) by the number n (more precisely, by the *numeral* for the number n, which we will blithely write as 'n', just as we will write '17', knowing that we really mean 'ssssssssssssssssss0'). This will yield a new formula, which we shall call 'G':

(G) $\quad\quad\quad \sim (\exists x)$ Dem $(x,$ Sub $(n, 17, n))$

This is the formula we promised. As it is a specialization of the formula on line (1), its meta-mathematical meaning is simply: 'The formula with Gödel number sub $(n, 17, n)$ is not demonstrable'. And now, as there are no (unquantified) variables left in it, G's meaning is definite.

The formula G occurs within PM, and therefore it must have a Gödel number, g. What is the value of g? A little reflection shows that $g =$ sub $(n, 17, n)$.[32] To

---

[32] Note the key distinction between the number itself and its formal counterpart inside PM. The former is sub $(n, 17, n)$, with lowercase 's', while the latter is the string we abbreviate as 'Sub $(n, 17, n)$', with uppercase 'S'. Otherwise put, 'sub $(n, 17, n)$' denotes an actual *quantity,* much as, say, the informal arithmetical expres-

see this, we need but recall that sub (n, 17, n) is the Gödel number of whatever formula results when we substitute n (or rather, its numeral) for the variable with Gödel number 17 (i.e., for 'y') inside the formula whose Gödel number is that same n itself. But the formula G was obtained in precisely that manner! That is, we started with the formula having Gödel number n; then we replaced all copies of 'y' in it by copies of the numeral for n. And so, sub (n, 17, n) is the Gödel number of G.

We must now recall that G is the mirror image *within* PM of the meta-mathematical statement: 'The formula with Gödel number g is not demonstrable'. It follows, then, that G represents, inside PM, the meta-mathematical statement: 'The formula G is not demonstrable'. In a word, the PM formula G can be construed as asserting of itself that it is not a theorem of PM.

(ii) We come to the second step—the proof that G is not, in fact, a theorem of PM. Gödel's argument showing this resembles the development of the Richard Paradox, but steers clear of its fallacious reasoning.[33] The argument is relatively unencumbered. It pro-

---

sion '2 × 5' denotes a *quantity* (namely, ten), whereas 'Sub (n, 17, n)' denotes a number-naming *string* inside PM, much like the number-naming string 'ss0 × sssss0'.

[33] It may be useful to make explicit the resemblance as well as the dissimilarity of the present argument to that of the Richard Paradox. The crux is that G is not identical with the meta-mathematical statement with which it is associated, but only *represents* (or mirrors) the latter within PM. In the Richard Paradox, the

ceeds by showing that *if* the formula G were demonstrable, then its formal negation (namely, the formula '($\exists x$) Dem (x, Sub (n, 17, n))', whose interpretation is 'There exists a demonstration of G inside PM') would also be demonstrable; and, conversely, that *if* the formal negation of G were demonstrable, then G itself would also be demonstrable. Thus we have: G is demonstrable if, and only if, ~G is demonstrable.[34] But

---

number n is the number associated with a certain *meta-mathematical* expression. In Gödel's construction, the number n is associated with a certain *formula* belonging to PM, and it is merely by happenstance, so to speak, that this formula represents a meta-mathematical statement. In the development of the Richard Paradox, the question is whether the number n possesses the *meta-mathematical* property of being Richardian. In the Gödel construction, the question is whether the number $g = \text{sub } (n, 17, n)$ possesses a certain *arithmetical* property—namely, the property that the assertion 'dem (x, g)' holds for no cardinal number x whatsoever. There is therefore no confusion, in the Gödel construction, between statements *within* PM and statements *about* PM, such as occurs in the Richard Paradox.

[34] This is not what Gödel actually proved; the statement in the text is an adaptation of a stronger result established by J. Barkley Rosser in 1936, and it is used for the sake of simplicity in exposition. What Gödel actually showed is that if G is demonstrable, then ~G is demonstrable (so that PM is inconsistent); while if ~G is demonstrable, then PM is ω-inconsistent.

What is ω-inconsistency? Let 'P' represent an arithmetical predicate. Then a formal calculus C is ω-inconsistent if it is possible to demonstrate inside C both the formula '($\exists x$) P(x)' (i.e., 'There is some number that has property P') and also each of the infinite set of formulas '~P(0)', '~P(s0)', '~P(ss0)', etc. (i.e., '0 does not have property P', '1 does not have property P', '2 does not have property

as we noted earlier, if a formula and its formal negation can both be derived in some formal calculus, then the calculus is not consistent. Turning this around, then, we reason that if PM is a consistent formal calculus, neither the formula G nor its negation can be demonstrable. In short, if PM is consistent, then G is necessarily formally undecidable.[35]

———

P', and so on). A little reflection shows that if C is inconsistent, it is also ω-inconsistent (for *all* strings are theorems in an inconsistent system); however, the converse does not necessarily hold: C may be ω-inconsistent without being inconsistent. In other words, not only '($\exists x$) $P(x)$' but also each of the family of formulas quoted above might be theorems of C while '~($\exists x$) $P(x)$' is a non-theorem, in which case C would be ω-inconsistent without being inconsistent.

It may seem absurd for each member of the family '~$P(0)$', '~$P(s0)$', etc., to be a theorem if '~($\exists x$) $P(x)$' is a non-theorem. After all, the family *collectively* asserts that no number has property P, while '~($\exists x$) $P(x)$' *singly* asserts that no number has property P. Does the latter not follow directly from the former? And how could '($\exists x$) $P(x)$', which asserts that *some* number has property P, be a theorem? Does it not directly contradict the family? Both worries seem justified if (like any human) you take *meanings* into account, but C is merely a formal calculus—only rules of inference, not meanings, are relevant. If some rule could take the whole family into account in one swoop, the worries would be justified—but although a rule can involve any *finite* number of formulas, no rule can cite an *infinite* number of formulas. (Recall Hilbert's insistence on *finitistic* procedures, in Chapter III.) And so this type of situation, queer though it is, can hold.

[35] We shall outline the first half of Gödel's argument: that if G is demonstrable, then ~G is demonstrable. Suppose G were demonstrable. This would mean that *there exists a sequence of formulas of PM constituting a proof for G.* We proceed to translate this metamathematical statement into a numerical one. Let the Gödel num-

(iii) This conclusion may not appear at first sight to be of capital importance. Why is it so remarkable, it may be asked, that an undecidable formula can be constructed within PM? There is a surprise in store that illuminates the profound implications of this result. For, although the formula G is undecidable if PM is consistent, it can nevertheless be shown by *meta-mathematical* reasoning that *G is true.* (Certainly either G or its negation ~G must be true, since they make two opposite claims about the world of numbers; one of these claims has to be true and the other one has to

---

ber of this hypothetical proof of G be k. Since the relationship dem (x, z) is the number-theoretical counterpart of "such-and-such is a proof of so-and-so", dem (x, z) must be true when the value of x is k and the value of z is the Gödel number of G. In other words, dem $(k, \text{sub } (n, 17, n))$ must be a fact of arithmetic. But given that dem (x, z) is a primitive recursive relationship (we agreed to take this on faith), its formal counterpart inside PM behaves properly, in the sense that 'Dem (sss . . . sss0, Sub (sss . . . sss0, sss . . . sss0, sss . . . sss0))' must be a theorem of PM, where the numbers of copies of 's' are, respectively, k, n, 17, and n. More concisely, 'Dem $(k, \text{Sub } (n, 17, n))$' must be a theorem. But, with the help of a rule of inference of PM, which says that from a theorem having the form 'P(k)' ('The number k has property P') one can derive the theorem '$(\exists x)$ P(x)' ('Some number has property P'), we can immediately derive the formula '$(\exists x)$ Dem $(x, \text{Sub } (n, 17, n))$'. But this is the formal negation of G. We have therefore shown that if the formula G is demonstrable, its formal negation is also demonstrable. It follows that if PM is consistent, G cannot be demonstrable within it.

A somewhat analogous but more complicated argument is required to show that if ~G is demonstrable, then PM is ω-inconsistent. We shall not attempt to outline it.

be false. The question is which one is right, and which one is wrong.)

It is quite straightforward to see that what G says is true. Indeed, as we observed earlier, G says: 'There is no PM demonstration of G'. (At least this is the *meta-mathematical* interpretation of G; when read at the number-theoretical level, G merely says that there is no number x that bears a certain relationship—namely, the 'dem' relationship—to the number sub $(n, 17, n)$. To convince ourselves that G is true, it suffices to consider only the former interpretation.) But we have just shown that *G is undecidable* within PM, so in particular *G has no proof* inside PM. But that, recall, is just what G asserts! So G asserts the truth. The reader should carefully note that we have established a number-theoretical truth not by deducing it formally from the axioms and rules of a formal system, but by a meta-mathematical argument.

(iv) We now remind the reader of the notion of "completeness" introduced in the discussion of the sentential calculus. It was explained there that a deductive system is said to be "complete" if every true statement that can be expressed in the system is formally deducible from the axioms by the rules of inference. If this is not the case—that is, if not every true statement expressible in the system is deducible—then the system is said to be "incomplete." But, since we have just established that G is a true formula of PM that is not formally deducible within PM, it follows that PM is an incomplete system—on the hypothesis, of course, that it is consistent.[36]

Moreover, PM is in even greater trouble than one might at first think, for it turns out to be not just incomplete, but *essentially* incomplete: even if G were added as a further axiom, the augmented system would still not suffice to yield formally *all* arithmetical truths. For, if the initial axioms were augmented in this manner, another true but undecidable formula could be constructed in the enlarged system. This formula would involve a slightly more complex number-theoretical relationship—dem$'$ (x, z), let us say—since the notion of "demonstrability" in the new system would be slightly more complex than it was in PM, as the new system has one extra axiom. The undecidable formula belonging to the new system is constructed merely by imitating the recipe by which Gödel specified a true but undecidable formula in PM itself. This method for producing undecidable formulas can be

---

[36] We could have reached this conclusion without going through the reasoning in step (iii) — that is, without knowing which one of G and ~G expresses a truth—since we had already concluded that G is undecidable, meaning that neither G nor ~G is demonstrable inside PM. Given that one of the two of these formulas *must* express a truth, and given that neither of them is demonstrable in PM, this in itself means that PM is incomplete, even if it leaves us unsure which of G and ~G is the culprit. Perhaps it is more comforting to know which of the two is the culprit, but it is not a necessary feature of the argument.

carried out no matter how often the initial system is enlarged. Nor does it depend in any crucial manner on peculiarities of Russell and Whitehead's formal cal-culus PM. The trick works no matter what system is taken as a starting point, as long as that system is totally formal and as long as it contains axioms setting forth the elementary properties of whole numbers, including addition and multiplication.

We are thus compelled to recognize a fundamental limitation concerning the power of formal axiomatic reasoning. Contrary to all prior belief, the vast conti-nent of arithmetical truth cannot be brought into sys-tematic order by laying down for once and for all a fixed set of axioms and rules of inference from which *every* true arithmetical statement can be formally de-rived. For anyone inclined to believe that the essence of mathematics is purely formal axiomatic reasoning, this must come as a shocking revelation.

(v) We come at last to the coda of Gödel's amazing intellectual symphony. The steps have been traced by which he grounded the meta-mathematical statement: 'If PM is consistent, it is incomplete'. But it can also be shown that this conditional statement *taken as a whole* is represented by a *demonstrable* formula within PM.

This crucial formula can be easily constructed. As we explained in Section V, the meta-mathematical state-ment 'PM is consistent' is equivalent to the claim 'There is at least one formula of PM that is not demon-strable inside PM'. By Gödel's mapping of meta-mathematics into the world of numbers, this corresponds to the number-theoretical claim 'There is at least one number y such that no x whatsoever bears the relationship dem to y'. In other words, 'Some number y has the property that for no x does the relationship dem (x, y) hold'. We can readily translate this into the formal notation of PM:

(A) $(\exists y) \sim (\exists x)$ Dem (x, y)

We can restate the meta-mathematical interpretation of A as follows: 'There is at least one formula [whose Gödel number is y] for which no proposed sequence of formulas [whose Gödel number is x] constitutes a proof inside PM'.

The formula A therefore represents the antecedent clause of the meta-mathematical statement 'If PM is consistent, it is incomplete'. On the other hand, the consequent clause in this statement—namely, 'It [PM] is incomplete'—is equivalent to saying, about any true but non-demonstrable formula X, 'X is not a theorem of PM'. Luckily, we know of one such formula X—namely, our old friend, the formula G. We can therefore translate the consequent clause into the formal language of PM by writing the string that says 'G is not a theorem of PM'. And it is none other than G itself that says this. And so G can be used as the consequent clause of our conditional meta-mathematical statement.

If we put it all together, then we arrive at the conclusion that the conditional sentence 'If PM is consistent, then it is incomplete' is represented inside PM by the formula:

$$(\exists y) \sim (\exists x) \text{ Dem } (x, y) \supset$$
$$\sim (\exists x) \text{ Dem } (x, \text{Sub } (n, 17, n))$$

which, for the sake of brevity, can by symbolized as 'A ⊃ G'. (This formula can be shown to be formally demonstrable inside PM, but we shall not in these pages undertake the task.)

We now show that the formula A is not demonstrable in PM. For suppose it were. Then, since the formula 'A ⊃ G' *is* demonstrable, by use of the Rule of Detachment (recall Chapter V), the formula G would be demonstrable. But, unless PM is inconsistent, G is formally undecidable—that is, it is not demonstrable. Thus, if PM is consistent, the formula A is not demonstrable in it.

Where does this lead us? The formula A is a formal expression inside PM of the meta-mathematical claim 'PM is consistent'. If, therefore, this meta-mathematical claim could be *informally* established by some chain of steps of reasoning, and if that chain of steps could be mapped onto a sequence of formulas constituting a proof in PM, then the formula A would itself be demonstrable inside PM. But this, as we have just seen, is impossible, if PM is consistent. The grand final step is thus before us: we are forced to conclude that *if* PM is consistent, its consistency cannot be established by any meta-mathematical reasoning that can be mirrored within PM itself!

This imposing result of Gödel's analysis should not be misunderstood: it does *not* exclude a meta-mathematical proof of the consistency of PM. What it excludes is a proof of consistency that can be mirrored inside PM.[37]

Meta-mathematical arguments establishing the consistency of formal systems such as PM have, in fact, been devised, notably by Gerhard Gentzen, a member of the Hilbert school, in 1936, and by others since then.[38] These proofs are of great logical significance,

---

[37] The reader may be helped on this point by the reminder that, similarly, the proof that it is impossible to trisect an arbitrary angle with compass and straight-edge does *not* mean that an angle cannot be trisected by any means whatsoever. To the contrary, an arbitrary angle can be trisected if, for example, in addition to the use of compass and straight-edge, one is permitted to employ a a fixed distance marked on the straight-edge.

[38] Gentzen's proof depends on arranging all the demonstrations inside PM in a linear order according to their degree of "simplicity" (note the resemblance to the formal version of the Richard Paradox, alluded to in the previous chapter). The arrangement turns out to have a pattern that is of a certain "transfinite ordinal" type. (The theory of transfinite ordinal numbers was created by the German mathematician Georg Cantor in the nineteenth century.) The proof of consistency is obtained by applying to this linear order a rule of inference called "the principle of transfinite induction." Gentzen's argument cannot be mirrored within PM. Moreover, although most specialists in mathematical logic do not question the cogency of the proof, it is not finitistic in the sense of Hilbert's original stipulations for an absolute proof of consistency.

among other reasons because they propose new forms of meta-mathematical constructions, and because they thereby help make clear how the class of rules of inference needs to be enlarged if the consistency of PM and related systems is to be established. But these proofs cannot be mirrored inside the systems that they concern, and, since they are not finitistic, they do not achieve the proclaimed objectives of Hilbert's original program.

VIII

Concluding Reflections

The import of Gödel's conclusions is far-reaching, though it has not yet been fully fathomed. These conclusions show that the prospect of finding for every deductive system (and, in particular, for a system in which the whole of number theory can be expressed) an absolute proof of consistency that satisfies the finitistic requirements of Hilbert's proposal, though not logically impossible, is most unlikely.[39] They show also that there are an endless number of true arithmetical statements which cannot be formally deduced from any given set of axioms by a closed set of rules of inference.

---

[39] The possibility of constructing a finitistic absolute proof of consistency for a formal system such as *Principia Mathematica* is not excluded by Gödel's results. Gödel showed that no such proof is possible that can be mirrored inside *Principia Mathematica*. His argument does not eliminate the possibility of strictly finitistic proofs that cannot be mirrored inside *Principia Mathematica*. But no one today appears to have a clear idea of what a finitistic proof would be like that is *not* capable of being mirrored inside *Principia Mathematica*.

It follows that an axiomatic approach to number theory cannot fully characterize the nature of number-theoretical truth. It follows, also, that what we understand by the process of mathematical proof does not coincide with the exploitation of a formalized axiomatic method. A formalized axiomatic procedure is based on an initially determined and fixed set of axioms and transformation rules. As Gödel's own arguments show, no antecedent limits can be placed on the inventiveness of mathematicians in devising new methods of proof. Consequently, no final account can be given of the precise nature of valid mathematical demonstrations. In the light of these circumstances, whether an all-inclusive definition of mathematical or logical truth can be devised, and whether, as Gödel himself appears to believe, only a thoroughgoing philosophical "realism" of the ancient Platonic type can supply an adequate definition, are problems still under debate and too difficult for further consideration here.[40]

––––––––

[40] Platonic realism takes the view that mathematics does not create or invent its "objects," but discovers them as Columbus discovered America. Now, if this is true, the objects must in some sense "exist" prior to their discovery. According to Platonic doctrine, the objects of mathematical study are not found in the spatio-temporal order. They are disembodied eternal Forms or Archetypes, which dwell in a distinctive realm accessible only to the intellect. On this view, the triangular or circular shapes of physical bodies that can be perceived by the senses are not the proper

Gödel's conclusions bear on the question whether a calculating machine can be constructed that would match the human brain in mathematical intelligence. Today's calculating machines have a fixed set of directives built into them; these directives correspond to the fixed rules of inference of formalized axiomatic procedure. The machines thus supply answers to problems by operating in a step-by-step manner, each step being controlled by the built-in directives. But, as Gödel showed in his incompleteness theorem, there are innumerable problems in elementary number theory that fall outside the scope of a fixed axiomatic method, and that such engines are incapable of answering, however intricate and ingenious their built-in mechanisms may be and however rapid their operations. Given a definite problem, a machine of this type might be built for solving it; but no one such machine can be built for solving every problem. The human brain may, to

———

objects of mathematics. These shapes are merely imperfect embodiments of an indivisible "perfect" Triangle or "perfect" Circle, which is uncreated, is never fully manifested by material things, and can be grasped solely by the exploring mind of the mathematician. Gödel appears to hold a similar view when he says, "Classes and concepts may . . . be conceived as real objects . . . existing independently of our definitions and constructions. It seems to me that the assumption of such objects is quite as legitimate as the assumption of physical bodies and there is quite as much reason to believe in their existence" (Kurt Gödel, "Russell's Mathematical Logic," in *The Philosophy of Bertrand Russell* (ed. Paul A. Schilpp, Evanston and Chicago, 1944), p. 137).

be sure, have built-in limitations of its own, and there may be mathematical problems it is incapable of solving. But, even so, the brain appears to embody a structure of rules of operation which is far more powerful than the structure of currently conceived artificial machines. There is no immediate prospect of replacing the human mind by robots.

Gödel's proof should not be construed as an invitation to despair or as an excuse for mystery-mongering. The discovery that there are number-theoretical truths which cannot be demonstrated formally does not mean that there are truths which are forever incapable of becoming known, or that a "mystic" intuition (radically different in kind and authority from what is generally operative in intellectual advances) must replace cogent proof. It does not mean, as a recent writer claims, that there are "ineluctable limits to human reason." It does mean that the resources of the human intellect have not been, and cannot be, fully formalized, and that new principles of demonstration forever await invention and discovery. We have seen that mathematical propositions which cannot be established by formal deduction from a given set of axioms may, nevertheless, be established by "informal" meta-mathematical reasoning. It would be irresponsible to claim that these formally indemonstrable truths established by meta-mathematical arguments are based on nothing better than bare appeals to intuition.

Nor do the inherent limitations of calculating machines imply that we cannot hope to explain living matter and human reason in physical and chemical terms. The possibility of such explanations is neither precluded nor affirmed by Gödel's incompleteness theorem. The theorem does indicate that the structure and power of the human mind are far more complex and subtle than any non-living machine yet envisaged. Gödel's own work is a remarkable example of such complexity and subtlety. It is an occasion, not for dejection, but for a renewed appreciation of the powers of creative reason.

Appendix

Notes

1. (page 10) It was not until 1899 that the arithmetic of cardinal numbers was axiomatized, by the Italian mathematician Giuseppe Peano. His axioms are five in number. They are formulated with the help of three undefined terms, acquaintance with the latter being assumed. The terms are: *'number'*, *'zero'*, and *'immediate successor of'*. Peano's axioms can be stated as follows:

1. Zero is a number.
2. The immediate successor of a number is a number.
3. Zero is not the immediate successor of a number.
4. No two numbers have the same immediate successor.
5. Any property belonging to zero, and also to the immediate successor of every number that has the property, belongs to all numbers.

The last axiom formulates what is often called the "principle of mathematical induction."

2. (page 39) The reader may be interested in seeing a fuller account than the text provides of the logical principles and rules of inference tacitly employed even in elementary mathematical demonstrations. We shall first analyze the reasoning that yields line 6 in Euclid's proof, from lines 3, 4, and 5.

We designate the letters '*p*', '*q*', and '*r*' as "sentential variables," because sentences may be substituted for them. Also, to economize space, we write conditional statements of the form 'if *p* then *q*' as '*p* ⊃ *q*' and we call the expression to the left of the horseshoe sign '⊃' the "antecedent," and the expression to the right of it the "consequent." Similarly, we write '*p* ∨ *q*' as short for the alternative form 'either *p* or *q*'.

There is a theorem in elementary logic which reads:

$$(p \supset r) \supset [(q \supset r) \supset ((p \lor q) \supset r)]$$

It can be shown that this formulates a *necessary truth*. The reader will recognize that this formula states more compactly what is conveyed by the following much longer statement:

If (if *p* then *r*), then [if (if *q* then *r*) then (if (either *p* or *q*) then *r*)]

As pointed out in the text, there is a rule of inference in logic called the Rule of Substitution for Sentential Variables. According to this Rule, a sentence S_2 follows logically from a sentence S_1 which contains sentential variables, if the former is obtained from the latter by uniformly substituting any sentences for the

variables. If we apply this rule to the theorem just mentioned, substituting '*y* is prime' for '*p*', '*y* is composite'
for '*q*', and '*x* is not the greatest prime' for '*r*', we
obtain the following:

(*y* is prime ⊃ *x* is not the greatest prime)
⊃ [(*y* is composite ⊃ *x* is not the greatest prime)
⊃ ((*y* is prime ∨ *y* is composite) ⊃ *x* is not the
greatest prime)]

The reader will readily note that the conditional
sentence within the first pair of parentheses (it occurs
on the first line of this instance of the theorem) simply
duplicates line 3 of Euclid's proof. Similarly, the conditional sentence within the first pair of parentheses
inside the square brackets (it occurs as the second line
of the instance of the theorem) duplicates line 4 of the
proof. Also, the alternative sentence inside the square
brackets duplicates line 5 of the proof.

We now make use of another rule of inference
known as the Rule of Detachment (or "Modus Po-
nens"). This rule permits us to infer a sentence S_2 and
from two other sentences, one of which is S_1 and the
other, $S_1 ⊃ S_2$. We apply this Rule three times: first,
using line 3 of Euclid's proof and the above instance
of the logical theorem; next, the result obtained by this
application and line 4 of the proof; and, finally, this
latest result of the application and line 5 of the proof.
The outcome is line 6 of the proof.

The derivation of line 6 from lines 3, 4, and 5 thus
involves the tacit use of two rules of inference and a

theorem of logic. The theorem and rules belong to the elementary part of logical theory, the sentential calculus. This deals with the logical relations between statements compounded out of other statements with the help of sentential connective, of which '⊃' and '∨' are examples. Another such connective is the conjunction 'and', for which the dot '·' is used as a shorthand form; thus the conjunctive statement '*p* and *q*' is written as '*p* · *q*'. The sign '∼' represents the negative particle 'not'; thus 'not-*p*' is written as '∼*p*'.

Let us examine the transition in Euclid's proof from line 6 to line 7. This step cannot be analyzed with the help of the sentential calculus alone. A rule of inference is required which belongs to a more advanced part of logical theory—namely, that which takes note of the internal complexity of statements embodying expressions such as 'all', 'every', 'some', and their synonyms. These are traditionally called *quantifiers,* and the branch of logical theory that discusses their role is the theory of quantification.

It is necessary to explain some of the notation employed in this more advanced sector of logic, as a preliminary to analyzing the transition in question. In addition to the sentential variables for which sentences may be substituted, we must consider the category of "individual variables," such as '*x*', '*y*', '*z*', etc., for which the names of individuals can be substituted. Using these variables, the universal statement 'All primes greater than 2 are odd' can be rendered: 'For every *x*, if *x* is a prime greater than 2, then *x* is odd'. The expression 'for every *x*' is called the *universal quantifier,* and in current logical notation is abbreviated by the sign '(*x*)'. The universal statement may therefore be written:

(*x*) (*x* is a prime greater than 2 ⊃ *x* is odd)

Furthermore, the "particular" (or "existential") statement 'Some integers are composite' can be rendered by 'There is at least one *x* such that *x* is an integer and *x* is composite'. The expression 'there is at least one *x*' is called the *existential quantifier,* and is currently abbreviated by the notation '(∃*x*)'. The existential statement just mentioned can be transcribed:

(∃*x*) (*x* is an integer · *x* is composite)

It is now to be observed that many statements implicitly use more than one quantifier, so that in exhibiting their true structure several quantifiers must appear. Before illustrating this point, let us adopt certain abbreviations for what are usually called predicate expressions or, more simply, predicates. We shall use 'Pr (*x*)' as short for '*x* is a prime number'; and 'Gr (*x, z*)' as short for '*x* is greater than *z*'. Consider the statement: '*x* is the greatest prime'. Its meaning can be made more explicit by the following locution: '*x* is a prime, and, for every *z* which is a prime but different from *x*, *x* is greater than *z*'. With the help of our various abbreviations, the statement '*x* is the greatest prime' can be written:

$$\text{Pr } (x) \cdot (z) \, [(\text{Pr } (z) \cdot \sim (x = z)) \supset \text{Gr } (x, z)]$$

Literally, this says: '*x* is a prime and, for every *z*, if *z* is a prime and *z* is not equal to *x* then *x* is greater than *z*'. We recognize in the symbolic sequence a formal, painfully explicit rendition of the content of line 1 in Euclid's proof.

Next, consider how to express in our notation the statement '*x* is not the greatest prime', which appears as line 6 of the proof. This can be presented as:

$$\text{Pr } (x) \cdot (\exists z) \, [\text{Pr } (z) \cdot \text{Gr } (z, x)]$$

Literally, it says: '*x* is a prime and there is at least one *z* such that *z* is a prime and *z* is greater than *x*'.

Finally, the conclusion of Euclid's proof, line 7, which asserts that there is no greatest prime, is symbolically transcribed by:

$$(x) \, [\text{Pr } (x) \supset (\exists z) \, (\text{Pr } (z) \cdot \text{Gr } (z, x))]$$

which says: 'For every *x*, if *x* is a prime, there is at least one *z* such that *z* is a prime and *z* is greater than *x*'. The reader will observe that Euclid's conclusion implicitly involves the use of more than one quantifier.

We are ready to discuss the step from Euclid's line 6 to line 7. There is a theorem in logic which reads:

$$(p \cdot q) \supset (p \supset q)$$

or when translated, 'If both *p* and *q*, then (if *p* then *q*)'. Using the Rule of Substitution, and substituting 'Pr (x)' for '*p*', and '$(\exists z) \, [\text{Pr } (z) \cdot \text{Gr } (z, x)]$' for '*q*', we obtain:

$$(\text{Pr } (x) \cdot (\exists z) \, [\text{Pr } (z) \cdot \text{Gr } (z, x)]) \supset$$
$$(\text{Pr } (x) \supset (\exists z) \, [\text{Pr } (z) \cdot \text{Gr } (z, x)])$$

The antecedent (first line) of this instance of the theorem simply duplicates line 6 of Euclid's proof; if we apply the Rule of Detachment, we get

$$(\text{Pr } (x) \supset (\exists z) \, [\text{Pr } (z) \cdot \text{Gr } (z, x)])$$

According to a Rule of Inference in the logical theory of quantification, a sentence S_2 having the form '(x) $(\ldots x \ldots)$' can always be inferred from a sentence S_1 having the form '$(\ldots x \ldots)$'. In other words, the sentence having the quantifier '(x)' as a prefix can be derived from the sentence that does not contain the prefix but is like the former in other respects. Applying this rule to the sentence last displayed, we have line 7 of Euclid's proof.

The moral of our story is that the proof of Euclid's theorem tacitly involves the use not only of theorems and rules of inference belonging to the sentential calculus, but also of a rule of inference in the theory of quantification.

3. (page 54) The careful reader may demur at this point, wondering something like this. The property of being a tautology has been defined in notions of truth and falsity. Yet these notions obviously involve a reference to something *outside* the formal calculus. Therefore, the procedure mentioned in the text in effect offers an *interpretation* of the calculus, by supplying a model for the system. This being so, the authors

have not done what they promised, namely, to define a property of formulas in terms of purely structural features of the formulas themselves. It seems that the difficulty noted in Section II of the text—that proofs of consistency which are based on models, and which argue from the truth of axioms to their consistency, merely shift the problem—has not, after all, been successfully outflanked. Why then call the proof "absolute" rather than relative?

The objection is well taken when directed against the exposition in the text. But we adopted this form so as not to overwhelm the reader unaccustomed to a highly abstract presentation resting on an intuitively opaque proof. Because more venturesome readers may wish to be exposed to the real thing, to see an unprettified definition that is not open to the criticisms in question, we shall supply it.

Remember that a formula of the calculus is either one of the letters used as sentential variables (we will call such formulas "elementary") or a compound of these letters, of the signs employed as sentential connective, and of the parentheses. We agree to place each elementary formula in one of two mutually exclusive and exhaustive classes K_1 and K_2. Formulas that are not elementary are placed in these classes pursuant to the following conventions:

i) A formula having the form $S_1 \lor S_2$ is placed in class K_2 if *both* S_1 and S_2 are in K_2; otherwise, it is placed in K_1.

ii) A formula having the form $S_1 \supset S_2$ is placed in K_2, if S_1 is in K_1 and S_2 is in K_2; otherwise, it is placed in K_1.

iii) A formula having the form $S_1 \cdot S_2$ is placed in K_1, if *both* S_1 and S_2 are in K_1; otherwise, it is placed in K_2.

iv) A formula having the form $\sim S$ is placed in K_2, if S is in K_1; otherwise, it is placed in K_1.

We then define the property of being tautologous: a formula is a tautology if, and only if, it falls in the class K_1 no matter in which of the two classes its elementary constituents are placed. It is clear that the property of being a tautology has now been described without using any model or interpretation for the system. We can discover whether or not a formula is a tautology simply by testing its structure by the above conventions.

Such an examination shows that each of the four axioms is a tautology. A convenient procedure is to construct a table that lists all the possible ways in which the elementary constituents of a given formula can be placed in the two classes. From this list we can determine, for each possibility, to which class the nonelementary component formulas of the given formula belong, and to which class the entire formula belongs. Take the first axiom. The table for it consists of three columns, each headed by one of the elementary or nonelementary component formulas of the axiom, as well as by the axiom itself. Under each heading is indicated the class to which the particular item belongs, for each of the possible assignments of the elementary constituents to the two classes. The table is as follows:

p	$(p \lor p)$	$(p \lor p) \supset p$
K_1	K_1	K_1
K_2	K_2	K_1

The first column mentions the possible ways of classifying the sole elementary constituent of the axiom. The second column assigns the indicated non-elementary component to a class, on the basis of convention (i). The last column assigns the axiom itself to a class, on the basis of convention (ii). The final column shows that the first axiom falls in class K_1, irrespective of the class in which its sole elementary constituent is placed. The axiom is therefore a tautology.

For the second axiom, the table is:

p	q	$(p \lor q)$	$p \supset (p \lor q)$
K_1	K_1	K_1	K_1
K_1	K_2	K_1	K_1
K_2	K_1	K_1	K_1
K_2	K_2	K_2	K_1

The first two columns list the four possible ways of classifying the two elementary constituents of the axiom. The second column assigns the non-elementary component to a class, on the basis of convention (i). The last column does this for the axiom, on the basis of convention (ii). The final column again shows that the second axiom falls in class K_1 for each of the four possible ways in which the elementary constituents can be classified. The axiom is therefore a tautology. In a similar way the remaining two axioms can be shown to be tautologies.

We shall also give the proof that the property of being a tautology is hereditary under the Rule of Detachment. (The proof that it is hereditary under the Rule of Substitution will be left to the reader.) Assume that any two formulas S_1 and $S_1 \supset S_2$ are both tautologies; we must show that in this case S_2 is a tautology. Suppose S_2 were not a tautology. Then, for at least one classification of its elementary constituents, S_2 will fall in K_2. But, by hypothesis, S_1 is a tautology, so that it will fall in K_1 for all classifications of its elementary constituents—and, in particular, for the classification which requires the placing of S_2 in K_2. Accordingly, for this latter classification, $S_1 \supset S_2$ must fall in K_2, because of the second convention. However, this contradicts the hypothesis that $S_1 \supset S_2$ is a tautology. In consequence, S_2 must be a tautology, on pain of this contradiction. The property of being a tautology is thus transmitted by the Rule of Detachment from the premises to the conclusion derivable from them by this rule.

One final comment on the definition of a tautology given in the text. The two classes K_1 and K_2 used in the present account may be construed as the classes of true and of false statements, respectively. But the account, as we have just seen, in no way depends on such an interpretation, even if the exposition is more easily grasped when the classes are understood in this way.

Brief Bibliography

CARNAP, RUDOLF   Logical Syntax of Language, New York, 1937.

FINDLAY, J.   "Goedelian sentences: a non-numerical approach," *Mind*, Vol. 51 (1942), pp. 259–265.

GÖDEL, KURT   "Über formal unentscheidbare Sätze der Principia Mathematica und verwandter Systeme I," *Monatshefte für Mathematik und Physik*, Vol. 38 (1931), pp. 173–198.

KLEENE, S. C.   Introduction to Metamathematics, New York, 1952.

LADRIÈRE, JEAN   Les Limitations Internes des Formalismes, Louvain and Paris, 1957.

MOSTOWSKI, A.   Sentences Undecidable in Formalized Arithmetic, Amsterdam, 1952.

QUINE, W. V. O.   Methods of Logic, New York, 1950.

ROSSER, J. BARKLEY   "An informal exposition of proofs of Gödel's theorems and Church's theorem," *Journal of Symbolic Logic*, Vol. 4 (1939), pp. 53–60.

TURING, A. M.   "Computing machinery and intelligence," *Mind*, Vol. 59 (1950), pp. 433–460.

WEYL, HERMANN   Philosophy of Mathematics and Natural Science, Princeton, 1949.

WILDER, R. L.   Introduction to the Foundations of Mathematics, New York, 1952.

Index